OVERSHOOT

OVERSHOOT

The Ecological Basis of Revolutionary Change

William R. Catton, Jr.

UNIVERSITY OF ILLINOIS PRESS
Urbana and Chicago

First Illinois paperback, 1982

© 1980 by the Board of Trustees of the University of Illinois

This book is printed on acid-free paper.

Library of Congress Cataloging-in-Publication Data Catton,
William Robert, 1926–

    Overshoot, the ecological basis of revolutionary change.
    Includes index.
    1. Human ecology. 2. Social change. I. Title
GF41.C37   303.4   80-13443
ISBN 0-252-00988-6 / 978-0-252-00988-4

Previously available in a cloth edition,
ISBN 0-252-00818-9.

to the memory of my father

Preface

In a future that is as unavoidable as it will be unwelcome, survival and sanity may depend upon our ability to cherish rather than to disparage the concept of human dignity. My purpose in writing this book has been to enhance that ability by providing a clear understanding of the ecological context of human life.

It is axiomatic that we are in no way protected from the consequences of our actions by remaining confused about the ecological meaning of our humanness, ignorant of ecological processes, and unmindful of the ecological aspects of history. I have tried to show the real nature of humanity's predicament not because understanding its nature will enable us to escape it, but because if we do not understand it we shall continue to act and react in ways that make it worse.

End-of-chapter notes are provided to document or clarify points about which readers may feel reasonable skepticism. A list of selected references is also provided at the end of each chapter to satisfy appetites for further knowledge which I trust the chapter content may arouse.

Acknowledgments

This book is dedicated to my father, for his awareness of the moral significance of the web of life whetted my concern and steadied my thinking. He lived life with fascination from 1895 to 1972—a period of momentous change and splendid but tragic human achievement. I wish it had been possible to discuss the contents of these pages with him; he would have understood, I'm sure, my reasons for writing. This book arose from thoughts engendered by a time of national shame and global troubles, when many Americans sought to work their way back to basic values. I have written with the conviction that my own experience gave me some advantage in comprehending the ecological foundations of our difficulties; my work, my family, the parts of the world I lived in, were such that I did not have to become preoccupied with the special grievances of one group toward another group. Thus I was not prevented from recognizing the similarity between the frustrations felt by each group and those experienced by its supposed adversaries. Moreover, my own exposure to population pressure, a major indicator of the common source of our mounting frustrations, has been sufficiently marginal and intermittent to permit me to see it in relief. Constant exposure to it would have prevented me (as it has prevented so many others) from seeing its real nature. Complete insulation from it would have precluded awareness and concern. Even with my advantageous situation, it took me years to see what I was looking at.

As is always the case, my eventual attainment of insight was immensely assisted by many other minds (including some I have never met in person, as well as many friends). In addition to the authors cited at the ends of the chapters or quoted at the beginnings of the six parts, various influences by the persons mentioned below are gratefully acknowledged.

It was when I sat in on lectures by my former colleague, Fred Campbell, now chairman of the department of sociology at the University of Washington, and engaged him in many conversations, that I came to see the extreme sociological cogency of ecological principles to which I had first begun to be sensitized in a non-academic but most enlightening context—wilderness areas in national parks.

As the ideas I have set forth in this book were coming into focus, I had the benefit of some very stimulating discussions with various individuals in the fields of forest ecology and management of wildland reserves. I should particularly acknowledge C. Frank Brockman, professor emeritus of forestry, and David R. M. Scott, professor of forestry, University of Washington, and a number of national park administrators in New Zealand, especially Mr. P. H. C. Lucas, then Director of National Parks and Reserves, Mr. Gordon Nicholls, Supervisor of National Parks, and Mr. J. H. Taylor, Assistant Supervisor for Interpretation. None of these persons, of course, should be assumed to agree fully with the direction my thought has since taken, or with my conclusions.

The undergraduate and graduate students in my classes in human ecology at universities in New Zealand and in the United States have helped my thinking become clearer by their patient endurance of its maturation, and by their challenging questions and counter-arguments.

Development of my understanding of the ecological paradigm has been enormously aided by the opportunity for unusually congenial and mutually enlightening collaboration with Riley E. Dunlap, associate professor of sociology at Washington State University, in the writing of a series of papers pertaining to environmental sociology.

I am indebted to the following persons for offering suggestions or critical comments on various portions of various versions of the manuscript (and again, there is no implication that any of them wholly concurs with my views): Charles E. Bowerman, professor emeritus of sociology, Washington State University; Robert E. L. Faris, professor emeritus of sociology, University of Washington; Dr. J. A. Gibb, Director, Ecology Division, Department of Scientific and Industrial Research, Wellington, New Zealand; George A. Knox, professor of zoology, University of Canterbury; Peggy Koopman-Boyden, senior lecturer in sociology, University of Canterbury; William R. Lassey, professor of rural sociology, Washington State University; Gerhard Lenski, professor of sociology, University of North Carolina; Armand L. Mauss, associate professor of sociology, Washington State University; Peter J. McKelvey, professor of forestry, University of Canterbury; J. Milton Yinger, professor of sociology, Oberlin College.

I am also indebted to John M. Wardwell, associate professor of sociology, Washington State University, and to my sons Stephen and Philip, for reading the entire manuscript in semi-final form, commenting critically and extensively, and encouraging me to finish it and publish it.

I would like to thank Ann Lowry Weir for her expert editing and Stewart L. Udall for contributing the Foreword.

Above all, I am abundantly grateful to my wife, Nancy. She not only continued to make life personally rewarding, even with the world in conspicuous and disheartening disarray, but she also read and re-read numerous drafts of chapter after chapter. She was enormously helpful to me by her continuing reluctance to accept ideas I could not put aside. She thereby kept me confronted with the tenaciously exuberant predispositions of citizens of a post-exuberant world—the audience I have sought to reach.

Foreword
by Stewart L. Udall

Limitless Expectations

We Americans came out of World War II infatuated with the miracles of science, firm in our faith that the same kinds of minds that had unlocked the secrets of the atom could solve any physical problem, and above all, convinced that the atomic age had solved once and for all any problem of future energy supplies.

A brief generation later, even after the Arab oil embargo had exposed our rapidly growing dependence on imported oil, we clung to the myth of the technological solution. It was called Project Independence—remember? Our scientists, we still fondly imagined, would develop the means to make America self-sufficient in energy (by 1980!).

From today's sober perspective, it may be hard to recall that the idea of unlimited resources, of "free" atomic energy that would eliminate resource shortages anywhere on earth by making it possible ultimately to synthesize whatever was needed, was universally and eagerly accepted by Americans at midcentury. Whatever was theoretically possible seemed technically feasible; just let the engineers get the bugs out. Whatever was technically feasible would ultimately also be economically affordable; because of cheap energy, unlimited economic expansion and global industrialization seemed to be rational goals. All this seemed to provide a conclusive rejoinder to the arguments of "Neo-Malthusian pessimists."

Belief that the limits to human activity had been or would soon be removed inspired exuberant predictions. We came to expect a flow of goods and machines and technical innovations that would lift standards of living everywhere. The business community believed that if sufficient funds for research and development were forthcoming, there were no limits to what technology could do. The one sure way to have a healthy economy seemed to be through rapid economic growth geared to modes of production which consumed larger and larger amounts of energy.

The first "Atoms for Peace" conference was convened in Geneva as 1955 ended, and bullish experts from the atomic-club countries laid out their blueprints for the expansion of nuclear power.

The first book that sought to provide a road map for the age of super-technology was *The Next Hundred Years*, prepared in 1957 by the noted geochemist, Dr. Harrison Brown, and some of his Cal Tech colleagues. It was the outgrowth of a series of seminars with executives of thirty of the nation's largest corporations. Impelled by the promise of inexhaustible energy, the authors provided this vision of what they called an emerging "technical-industrial civilization":

> If we are able in the decades ahead to avoid thermonuclear war, and if the present underdeveloped areas of the world are able to carry out successful industrialization programs, we shall approach the time when the world will be completely industrialized. And as we continue along this path we shall process ores of continually lower grade, until we finally sustain ourselves with materials obtained from the rocks of the earth's crust, the gases of the air, and the waters of the seas.

> By that time the mining industry as such will long since have disappeared and will have been replaced by vast, integrated multipurpose chemical plants supplied by rock, air, and sea water, from which will flow a multiplicity of products, ranging from fresh water to electric power, liquid fuels and metals.

It was believed that the main obstacle to the attainment of this resource El Dorado was not a shortage of resources (nor the threat of overpopulation), but a lack of enough trained scientists and engineers to build and maintain the technical wonderworks needed by the developing countries.

A further expression of what someone later called "the golden optimism of the 1950s" was contained in the Rockefeller Panel reports published in 1959 and 1960. These studies were prepared with guidance from men like Dean Rusk, Henry Kissinger, and Arthur Burns, and were treated by the press as unofficial White Papers on the American future. The Rockefeller experts endorsed the super-technology hypothesis in these words: "New technologies, more efficient extraction processes, new uses may open up new worlds. Even now we can discern the outlines of a future in which, through the use of the split atom, our resources of both power and raw materials will be limitless. . . ."

Policy of Aiming Too High

In 1962 a special committee of the National Academy of Sciences completed a report for President Kennedy containing recommendations for a new natural resources policy. This report, in sanguine tones, called for a basic reorientation of the U.S. approach to resources, and asserted that new breakthroughs in science had given mankind the potential to provide "dramatic increases" in energy and food. The breeder reactor was treated as something already accomplished. Mastery of controlled fusion was considered "probable" in a ten- to thirty-year period. The panel counselled the United States to shift away from a philosophy of conserving scarce resources (recommended a decade earlier by the Paley Commission to President Truman) to a policy described as "the wise management of plenty."

It was a landmark study, and its recommendations were unanimous. It cemented the consensus about technology and implied that if we ran out of petroleum or iron ore—or any other mineral—technology would soon come forth with a better, cheaper substitute. Such findings had a powerful influence on our national leaders and their perceptions. This influence was evident in the ebullient mood—and extravagant aims—of the period we later called the "Soaring '60s."

We aimed too high because we relied on the assurances of the technologists that we were living in an age in which there were "no problems, only solutions." This belief in an omnipotent science shaped efforts and expectations, both in Washington and in the country as a whole.

My Interior Department mirrored the general optimism about technology. One of Interior's principal research efforts in the 1960s was water desalting; we were certain that sooner or later a technical breakthrough would make it feasible to turn the world's deserts into irrigated gardens. The programs of the Interior Department were, however, a mere sideshow. Everyone, it seemed, had an eye-catching act. The physicists and engineers at the Atomic Energy Commission spent hundreds of millions of dollars working on pet "Plowshare" schemes to use atomic explosions to dig harbors, to increase the output of natural gas fields, and to prepare for the excavation of a new Panama Canal. Aerospace engineers and Pentagon weapons managers offered to demonstrate how their systems analysis method could be used to solve the problems of the cities.

But the high-wire performance in the tent of big technology was the Apollo program. Each successful flight was watched by a vast audience, and each exploit seemed to certify that American technology was all-powerful. In restrospect, we can already see that these flights became more than brilliant exhibitions of engineering. They were voyages of the imagination, voyages which led us on a joy ride of extravagant expectations.

It is easy to fix the exact date when our euphoria reached a zenith. It was the July week in 1969 when the astronauts walked on the moon. We celebrated this triumph with a mixture of awe and self-congratulation. President Nixon proclaimed that it was "the greatest week since the creation of the earth." A NASA official opined that the feat demonstrated we were "masters of the universe." "This proves that we can do whatever we decide to do," Americans concluded from this climax event.

But the 1970s were not so hospitable to big technology, or to the proponents of the mind-of-man doctrine. Indeed, they were a decade of disillusionment, of broken promises and unexpected developments. The Vietnam outcome was a setback to those who assumed that technical superiority in weapons would ensure victory over an agrarian nation. The reappearance of famines in Africa and Asia reminded us that science had not yet found ways to feed a fast-growing world. The oil embargo and the natural gas shortage shocked us into realizing that we still had no substitutes for our dwindling natural reserves. It was a symptom of the general disillusionment that the prideful, "We can do whatever we decide to do" of 1969 was replaced with the querulous lament, "But if we could put men on the moon, why can't we . . . ?"

These developments suggest that a major reorientation is necessary if we are to cope with an energy predicament which threatens to cripple our country. If the atomic age now appears to have been an age of overestimation, then it is vital that we put technology in perspective and gain a better understanding of its strengths, and its limits. This will entail a reassessment of the contributions of technology—and of cheap petroleum—to postwar progress. Such an inquiry might challenge, for example, the claim that the Green Revolution is a monumental triumph of science and technology. How much of our farm productivity increase has been due to science, and how much to cheap oil, superior U.S. soils, and the beneficent weather of recent years?

It is undeniable that scientific advances in agronomy, in plant genetics, in weed and pest control, and in new insights about the applications of fertilizers and the mechanization of farming have added substantially to agricultural output. But petroleum that was deceptively cheap has played a major role in nurturing this illusion of perpetually expandable abundance. It has supplied fuel for water pumps and processing plants and field machines, and it has served as low-cost raw material for fertilizers, herbicides, and insecticides. Studies have already shown that the food and fiber industry is the single largest user of energy in this country, so we should abandon the pretense that our success in farming is solely an achievement of science and technology.

It is equally instructive to take a fresh look at transportation, another sector supposedly the scene of engineering's greatest achievements. Again, the available evidence suggests that the contributions of technology have been overstated, while the role of cheap oil has been understated. Beginning with Henry Ford and the Wright brothers, technical mastery has been part of the American achievement in transportation. There have been remarkable accomplishments in the design and manufacturing of autos, trucks, airplanes, and rockets. But how much of this would have been possible if the United States had not been an oil-rich country? Could we have created huge, sprawling cities and an auto-centered transportation system if we had not been lavishly endowed with low-cost fuel?

It was not just the advances in pipeline technology that made it possible for people in New York and New Jersey to burn Texas gas in their homes; it was a one-time abundance of natural gas. Nor was it just the skill of aerospace engineers which allowed our air transport to flourish; dirt-cheap petroleum made the whole thing "fly."

Return from an Ego Trip

America has preened itself for three decades on the wizardry of its technologists. All the evidence suggests that we have consistently exaggerated the contributions of technological genius and underestimated the contributions of natural resources.

Myths die hard, but the evidence of our overshoot is accumulating daily. As President Carter discovered, it is not easy to take a country conditioned to believe that every problem has a technical solution and to persuade its citizens that a major change of orientation has become necessary.

Yet one thing is now obvious. To accept the hard path of belt-tightening and sacrifices, we must first trim back our technological optimism. We need, in short, something we lost in our haste to remake the world: a sense of limits, an awareness of the importance of earth's resources.

Perhaps at no time in human history has there been a more compelling need to re-examine public assumptions and to change national
expectations. *Overshoot* is a book that contributes to this vital task. As
William Catton shows, the long experience of mankind has included
a number of successful efforts to enlarge this planet's human carrying
capacity. Conventional history has not enabled us to sense the real
significance of this pattern or the real nature of the predicament we
now face. Unfortunately, history has not been written in ecological
language. Neither the writers nor the readers of history have been in
the habit of thinking about "carrying capacity," "niches," "symbiosis,"
or "overshoot."

The earlier enlargements of human carrying capacity were
achieved, as Catton makes clear very early, by *displacing* other species. Human tribes took over for human use portions of the life-supporting potential of the biosphere that would otherwise have sustained
other forms of life. That displacement process couldn't continue forever; eventually we had to run out of displaceable competitors. Continued takeover efforts then would have been at the expense of plant
and animal types upon which we (knowingly or not) were symbiotically dependent. That is one reason why recent efforts by our government (and by other nations, and international agencies) to protect endangered species have been more than mere indulgence in esthetic
hypersensitivity.

As the reader of this book will see, man's most recent expansions
of human niches have been achieved by quite a different method.
Rather than displacing competitor species, we have been using ever
more powerful technology to draw down geological reservoirs of energy and materials that do not get renewed in a seasonal cycle of organic growth. For a while, this draw-down permitted those lucky human beings in the most "developed" societies to "live it up." Catton
shows why the exuberant expansion of human numbers and human
activities to take advantage of necessarily temporary increases in human niches was a tragic mode of progress, and why such a technol-
ogy-based ego trip now means that the future holds for us not merely
a regrettable leveling-off of economic growth, but an institution-
threatening prospect of some deindustrialization and some decline of
population. He brings into sharp focus the seductive characteristics of
past technological breakthroughs that enticed mankind into this trap.
He shows how our technological servants have now become our competitors—competing with us for limited supplies of breathable air and
drinkable water. He spells out the consequences that have befallen
the populations of other species that overshot their habitat's carrying

capacity, provides evidence for concluding that this is the nature of the predicament we humans now confront, and makes it clear that we can no longer rely glibly on human "uniqueness" to bail us out. Human uniqueness, which has enabled us to have such colossal environmental impact, has become part of the problem, rather than a certain remedy.

Some of us will continue to hope that the outlook for humanity may prove less dismal than Catton discerns it to be. Those trained to work toward partial solutions to problems that may seem insoluble will—everyone should hope—continue their work. But even the most incorrigible optimists can benefit from reading this call to face the future realistically. Catton shows that, even in our present impasse, this can be done with determination. He insists that a realistic awareness of our ecological predicament and our own nature, together with courage based upon a genuinely appreciative recollection of our heritage, are the best means of equipping ourselves to adapt to the uncertain future brought upon us by our past "successes."

Contents

I | The Unfathomed Predicament of Mankind

Attention . . . the mental attitude which takes note of the outside world and manipulates it . . . is associated with habit on the one hand and with crisis on the other. When the habits are running smoothly the attention is relaxed; it is not at work. But when something happens to disturb the run of habit the attention is called into play and . . . establishes new and adequate habits, or it is its function to do so.

> — William Isaac Thomas
> *Sourcebook of Social Origins*, p. 17

Social theorists today work within a crumbling social matrix. . . . The old order has the picks of a hundred rebellions thrust into its hide.

> — Alvin W. Gouldner
> *The Coming Crisis of Western Sociology*, p. vii

. . . man is generally confronted with a rapidly growing condition of stress upon his environment that threatens his welfare and even his survival, and . . . popular attitudes and public institutions are not generally prepared to cope with this circumstance.

> — Lynton K. Caldwell
> *Man and His Environment: Policy and Administration*, p. xi

Signs of stress on the world's principal biological systems—forests, fisheries, grasslands, and croplands—indicate that in many places these systems have already reached the breaking point. Expecting these systems to withstand a tripling or quadrupling of population pressures defies ecological reality.

> — Lester R. Brown
> *The Twenty-ninth Day*, p. 92

The significance of crises is the indication they provide that an occasion for retooling has arrived.

> — Thomas S. Kuhn
> *The Structure of Scientific Revolutions*, p. 76

1 | Our Need for a New Perspective

Competition across Time

On the banks of the Volga in 1921 a refugee community was visited by an American newspaper correspondent who had come to write about the Russian famine.[1] Almost half the people in this community were already dead of starvation. The death rate was rising. Those still surviving had no real prospect of prolonged longevity. In an adjacent field, a lone soldier was guarding a huge mound of sacks full of grain. The American newsman asked the white-bearded leader of the community why his people did not overpower this one guard, take over the grain, and relieve their hunger. The dignified old Russian explained that the sacks contained *seed* to be planted for the next growing season. "We do not steal from the future," he said.

Today mankind is locked into stealing ravenously from the future. That is what this book is about. It is not just a book about famine or hunger. Famine in the modern world must be read as one of several symptoms reflecting a deeper malady in the human condition—namely, diachronic competition, a relationship whereby contemporary well-being is achieved at the expense of our descendants. By our sheer numbers, by the state of our technological development, and by being oblivious to differences between a method that achieved lasting increments of human carrying capacity and one that achieves only temporary supplements, we have made satisfaction of today's human aspirations dependent upon massive deprivation for posterity.

People of one generation have become indirect and unwitting antagonists of subsequent generations. Yet diagnoses of our plight—even ecological analyses—have not made clear one essential point. A major aim of this book is to show that commonly proposed "solutions" for problems confronting mankind are actually going to aggravate those problems. Proposed remedies for various parts of our predicament need to be evaluated by asking whether they will intensify the adversary relationship between people living today and people of the next generation, and the next . . .

The overlooked differences between methods that permanently enlarged human carrying capacity and more recent methods that have only enabled us temporarily to *evade* the world's limits can be seen if we contrast the way people *used to* seek the good life versus today's substitute expedient. In the mid-nineteenth century, it was "Go west, young man, and grow up with the country"—i.e., go where there is new land to take over, and use such an increment of carrying capacity to prosper. At the start of America's third century, however, it was "Try to speed up the economy"—i.e., try to draw down the finite reservoir of exhaustible resources a bit faster.

Because this book is meant to overcome our habit of mistaking techniques that evade limits for techniques that raise them, it is, in a sense, a book about how to read the news perceptively in revolutionary times. That cannot be done without certain unfamiliar but increasingly indispensable concepts. "Carrying capacity" is one of them. Until recently, only a few people outside such occupations as wildlife management or sheep and cattle ranching have even known this phrase. Its vital importance to all of us has not been as obvious as it is now becoming. The time has come for scholars and everyone else to take a piercing look at the relationship between the earth's *changing* capacity to support human inhabitants and the changing load imposed by our numbers and our requirements. The direction of recent change makes this relationship just about the most important topic there is for people to know about, and think about. We have come to the end of the time when it didn't seem to matter that almost no one saw the difference between ways of enlarging human carrying capacity and ways of exceeding it.

It has now become essential to recognize that all creatures, human or otherwise, impose a load upon their environment's ability to supply what they need and to absorb and transform what they excrete or discard. An environment's carrying capacity for a given kind of creature (living a given way of life) is the *maximum persistently feasible load*—just short of the load that would damage that environment's ability to support life of that kind. Carrying capacity can be expressed quantitatively as the number of us, living in a given manner, which a given environment can support *indefinitely*.

When the load at a particular time happens to be appreciably less than the carrying capacity, there is room for expansion of numbers, for enhancement of living standards, or both. If the load increases until it exceeds carrying capacity, overuse of the environment *reduces* its carrying capacity.[2] That is why it has become important to recognize the difference between increasing the number an environment can support indefinitely and surpassing that number by "accepting" environmental damage. Overuse of an environment sets up forces that will necessarily, in time, reduce the load to match the shrinkage of carrying capacity.

As these points begin to indicate, in order really to understand our future we need a clear-headed ecological interpretation of history, because the pressure of our numbers and technology upon manifestly limited resources has already put out of reach the previously acceptable solutions to many of our problems.[3] There remains steadfast resistance to admitting this, but facts are not repealed by refusal to face them. On the other hand, even the "alarmists" who have been warning of grave perils besetting mankind have not fathomed our present predicament.

I speak of "predicament," not "crisis," because I refer to conditions that are not of recent origin and will not soon abate.

In brief, that predicament and its background can be outlined as follows: Human beings, in two million years of cultural evolution, have several times succeeded in taking over additional portions of the earth's total life-supporting capacity, at the expense of other creatures. Each time, human population has increased. But man has now learned to rely on a technology that augments human carrying capacity in a necessarily temporary way—as temporary as the extension of life by eating the seeds needed to grow next year's food. Human population, organized into industrial societies and blind to the temporariness of carrying capacity supplements based on exhaustible resource dependence, responded by increasing more exuberantly than ever, even though this meant overshooting the number our planet could permanently support.[4] Something akin to bankruptcy was the inevitable sequel.

Old Ideas, New Situation

The sequel has begun to happen, but it is not generally recognized for what it is.[5] We have come to a time when old assumptions that compel us to misunderstand what is happening to us have to be abandoned.[6]

We and our immediate ancestors lived through an age of exuberant growth, overshooting permanent carrying capacity without knowing what we were doing. The past four centuries of magnificent progress were made possible by two non-repeatable achievements: (a)

5

discovery of a second hemisphere, and (b) development of ways to exploit the planet's energy savings deposits, the fossil fuels. The resulting opportunities for economic and demographic exuberance convinced people that it was natural for the future to be better than the past. For a while that belief was a workable premise for our lives and institutions. But when the New World became more populated than the Old World had been, and when resource depletion became significant, the future had to be seen through different lenses.

Assumptions that were once viable but have become obsolete must be replaced with a new perspective, one that enables us to see more effectively and to understand more accurately. This book seeks to articulate that needed perspective. It is no easy task, for the new way of seeing must differ sharply from traditional assumptions. Being unfamiliar, the new perspective will initially be distasteful and seem implausible. We shall continue to wish that some of the experiences it enables us to understand more clearly were not happening. But if we have the wisdom implied by the name we gave our own species, we must face the fact that continued misunderstanding of unwelcome experiences cannot prevent them from happening and cannot insulate us from their consequences.

People accustomed to expectations of magnificent progress have been appalled to find that they have lost their confidence in the future. The idea that mankind could encounter hardships that simply will not go away was unthinkable in the Age of Exuberance. This idea must be faced in the post-exuberant world. It seemed at last that it might really be faced when the thirty-ninth president of the United States decided (shortly after taking office) to emphasize energy *conservation* in response to manifest depletion of once-abundant fuels, rather than resorting to the traditional American urge to "produce our way out" of mounting difficulties. Important options had been lost irretrievably when humanity irrupted beyond the earth's permanent carrying capacity.[7] New and different imperatives now must be faced. Their ecological basis must be seen.

Man is like every other species in being able to reproduce beyond the carrying capacity of any finite habitat. Man is like no other species in that he is capable of thinking about this fact and discovering its consequences. Thinking about other species, man has seen their dependence upon environmental chemistry and upon the energy of sunlight. Man has recognized the many-faceted interdependence of diverse organisms, their impact upon their habitat, their impermanence, and their inability to foresee and evade the processes leading to their own displacement by successors. Thinking about our own species, however, at least until April, 1977, too many of us imagined ourselves exempt, supernatural. Until a president not yet worn down by the compromises inherent in officeholding nudged Congress and the American people into serious discussions of conservation, men and women throughout the United States and many other lands relied on technological progress to cure the very afflictions it had been causing.

Nature is going to require reduction of human dominance over the world ecosystem.[8] The changes this will entail are so revolutionary that we will be almost overwhelmingly tempted instead to prolong and augment our dominance at all costs. And, as we shall see, the costs will be prodigious. We are likely to do many things that will make a bad situation worse. It is hoped that the kind of enlightenment offered in this book may help curtail such tendencies.

The paramount need of post-exuberant humanity is to remain human in the face of dehumanizing pressures. To do this we must learn somehow to base exuberance of spirit upon something more lasting than the expansive living that sustained it in the recent past. But, as if we were driving a car that has become stuck on a muddy road, we feel an urge to bear down harder than ever on the accelerator and to spin our wheels vigorously in an effort to power ourselves out of the quagmire. This reflex will only dig us in deeper. We have arrived at a point in history where counter-intuitive thoughtways are essential.

Plan of the Book

To answer that need, the six parts of this book are intended as integrated contributions to an overdue "paradigm shift." The phrase is from Thomas Kuhn, who recognized that, even in the sciences, fundamental ideas about the way things work may *guide* our seeing (rather than simply *resulting from* what we have seen). A paradigm is an underlying basic idea of what kind of thing it is we are trying to understand. Such an idea of the nature of our subject matter prompts us to make certain kinds of inquiries; it also dissuades us from asking various other sorts of questions (or from seeing that we could, or that they would be relevant). So, when we *need* to ask questions made inconceivable by our old paradigm, we have come to need a new paradigm.

We get some of our fundamental ideas about what things are like and how they work from the kind of language we use in talking about them. As the philosopher Max Black has made clear, the vocabulary we are accustomed to using for describing what happens in the world around us will cause some aspects of reality to be emphasized and other aspects to be neglected. New words can yield new sensitivity, for vocabulary filters experience, shapes perception, and guides understanding.

This book uses an ecological vocabulary in describing and accounting for events that most people have been accustomed to thinking about in quite non-ecological terms. Accordingly, following the present chapter's outline of our topic, the three chapters in Part II should provide a truly eye-opening experience. Chapter 2 describes the process by which *Homo sapiens* painted himself into a corner. Our conventional, pre-ecological paradigm has prevented us from seeing that that is what we were doing. Chapter 3 shows how woefully misleading are some assumptions bred into us by the culture of exuberance. Chapter 4 outlines the alternative responses of various people to the obsolescence of traditionally exuberant expectations.

Then Part III (Chapter 5) highlights the contrast between the venerable American dream and its sequel. It examines some of the events of America's transformation from a vibrant young nation epitomizing the world's high hopes, into a country fumbling (with the rest of mankind) to come to terms with post-exuberant circumstances. It is especially intended to suggest that our persistent preoccupation with merely political facets of these events blinds us to (but does not protect us from) the enormity of the change.

The five chapters of Part IV spell out the consequences of our unwisdom in mistaking a temporary and unreliable increment of carrying capacity for a permanent expansion of opportunity and a durable form of progress. Chapter 6 sets forth basic principles of ecology, principles which have become as essential as literacy but which are habitually ignored by most leaders. Chapters 7 and 8 show how applicable ecological principles are to man. Chapter 9 corrects the misinterpretation of human uniqueness that has provided the excuse for ignoring these principles. Chapter 10 describes and explains the fateful course upon which we embarked when (understandably, but invalidly) we claimed independence from nature.

In Part V, Chapter 11 describes some common tactics of mental evasion by which people of the post-exuberant world have sought to say their extravagant dreams can yet be fulfilled. Chapter 12 shows how, as the world became post-exuberant, more and more of its people felt the kinds of pressure that dehumanize. Intensified intraspecific competition was undermining hope and decency. Typical responses to pressure aggravated the pressure.

But, as we see in Part VI, the future that was widely thought to be remote and even improbable was already arriving. Events that can be seen from the new ecological perspective to have been ominous previews of our impending de-civilization were not seen that way, as Chapter 13 shows, because of the old blinders we wore. Chapter 14 describes events that began at last to remove the blinders. The final chapter shows that mankind must gamble on an uncertain future, for phenomenally high stakes—and that, ironically, the less optimistic the assumptions we let ourselves make about the human prospect, the greater our chances of minimizing future hardships for our species. To keep from dehumanizing ourselves (and even gravitating toward genocide), we must stop demanding perpetual progress.

Of necessity, unfamiliar words are involved in the presentation of an unfamiliar worldview. A glossary at the end of the book is meant to minimize the inconvenience this terminology might otherwise cause.

Needed Realism

We live in a world where it is becoming increasingly evident that, for quite non-political reasons, governments and politicians cannot achieve the paradise they have habitually promised. Political habits persist, though, and people have seen their leaders continuing to dangle before them the kinds of carrots ordinary men and women know are becoming unattainable. The result has been erosion of faith in political processes. As that faith disintegrates, societies all over the world are floundering, or becoming dictatorial.

Even among Americans, who confront worldwide shrinkage of political liberty with a proud memory of two centuries' experimentation in democratic nation-building, faith in democracy has been seriously strained.[9] That strain may be reduced when politicians are astute enough to discover, and realistic enough to point out to their constituents, the non-political reasons why certain traditional goals are no longer attainable.

The alternative to chaos is to abandon the illusion that all things are possible. Mankind has learned to manipulate many of nature's forces, but neither as individuals nor as organized societies can human beings attain outright omnipotence. Many of us remain, to this day, beneficiaries of the once–New World's myth of limitlessness. But

9

circumstances have ceased to be as they were when that myth made sense. Unless we discard our belief in limitlessness, all of us are in danger of becoming its victims.

In today's world it is imperative that all of us learn the following core principle:

> Human society is inextricably part of a global biotic community, and in that community human dominance has had and is having self-destructive consequences.

That principle is the basis of the non-political obstacles that now frustrate human societal aspirations everywhere on earth. The obstacles are non-political because they are not uniquely human. In less-than-global ecosystems, as we shall see in Part IV, other dominant species before us have also been unavoidably self-destructive. We and our leaders need to understand such examples and to avail ourselves of the light they cast on our own history and our own situation. They exemplify a stark fact about life which we desperately need to grasp—and we need to see that grasping it will actually help us to adjust sanely to an unwelcome but inescapable future.

Because we haven't known these things, we, the human species, are inexorably tightening the two jaws of a vise around our fragile civilization. As we shall see in the next two chapters, there are already more human beings alive than the world's *renewable* resources can perpetually support. We have built complex societies that therefore *depend* on rapid use of exhaustible resources. Depletion of resources we don't know how to do without is *reducing* this finite planet's carrying capacity for our species. That is one jaw of the closing vise. The other is the accumulation of harmful substances that are unavoidably created by our life processes. There are so many of us, using so much technology, that these substances accumulate too fast for the global ecosystem to reprocess them; in fact, by overloading the natural reprocessing systems we are even breaking down their already limited capacity to set things right for us.[10] This accumulation of toxic materials also reduces the earth's human carrying capacity.

Oblivious of the fact that we were living between these two jaws, most of mankind has until recently applauded each turn of the vise handle. From the archaic perspective we are struggling to outgrow, it seemed like welcome progress. Hundreds of millions want the handle turned at least one or two more revolutions, for they have not yet fully shared in that progress. And that is what their leaders continue to promise.

Futile Vilification

Homo sapiens has not been the first type of organism to experience this vise-tightening, nor even the first species to *inflict upon itself* this kind of fate. Pre-human instances of this common phenomenon hold important lessons for us, as we shall see. For mankind, as the pressure intensifies, ignorance of its most fundamental causes (and ignorance even of how common the phenomenon has been in nature) makes it easy to succumb to the temptation to vilify particular human groups and individuals. "If only . . . ," we have been tempted to exclaim, " . . . if only those ———— weren't up to their nefarious business." Then history could resume its march of millennial progress (we suppose). "They" are the obstacles to our attainment of benevolent goals.

Depending upon which outraged in-group was doing the finger-pointing, "they" has referred to different out-groups. In place of "those ————," devout Maoist Chinese could read "Russian revisionists." Bedeviled Israelis could read "PLO terrorists," while embittered Arabs could read "land-grabbing Zionists." Angry Irish Catholics could read "Protestant extremists," and irate Ulster Protestants could read "IRA Provisionals." Black Rhodesians could read "the colonialist white minority regime"; Fidel Castro in Cuba could read "Yankee imperialists"; American motorists annoyed by increased gasoline prices could read "the oil cartel"; etc.

While vilification often brings emotional gratification, it brings no solution to our *common* plight. Indeed, it aggravates life's difficulties. Our common plight is not really due to villains. Too few of a troubled world's proliferating antagonists have known the concepts that would enable them to see the common roots of their own and their supposed adversaries' deprivations. Under pressure, people retreat from the mutual understanding mankind has so falteringly achieved. Pressure also makes us disinclined to comprehend the human relevance of nature's impersonal mechanisms. It behooves some who have borne the pressure only marginally to discern and discuss its nature, that all may stand some chance of abstaining from the plight-worsening actions to which pressure so easily tempts us.

The pressure to which we have collectively subjected ourselves, needs to be turned off, of course. But it is already too late to evade the future by so doing. This book is not a belated harangue for more birth control, much less a plea for one more revolution, or one last orgy of repentance. There is no point to another morbid wringing of hands over mankind's alleged "greed" or immoral myopia. Merely to deplore human appetites and short-sightedness is useless, without some effort to *understand* them (and to put them in perspective by comparing them with those of other species).

The purpose of this book, therefore, is to illuminate the nature and causes of the human predicament, so as to make possible some mitigation of its social, emotional, and moral effects. To mitigate the effects of post-exuberant pressures, we must recognize their deepest roots. We must learn to relate personally to what may be called "the ecological facts of life." We must see that those facts are affecting our lives far more importantly and permanently than the events that make headlines. To understand the human predicament now requires a truly ecological perspective.

We need an ecological worldview; noble intentions and a modicum of ecological information will not suffice. This new paradigm will help us see how the self-destructive ways of mankind are in many respects typical of the paths followed by other creatures. Until the reader has made the paradigm shift herein called for, it will doubtless be difficult to see why it matters whether we recognize the typicality of our seemingly unprecedented predicament. It will not be easy to see how finding our plight to be natural or ineluctable provides any consolation. As Kuhn pointed out, it is difficult (even in scientific discussions) for adherents of one paradigm to communicate effectively with those who perceive and reason in terms of a different paradigm. But I have tried to achieve communication across the paradigm barrier by writing as clearly and persuasively as I can, albeit from an unconventional perspective. I ask the reader to make an equally earnest *effort to achieve* the reorientation of thought represented by the following chapters. This is not a book to be read either casually or passively.

With active effort, the reader may find it possible to accept the view that, since human beings are not the first creatures to foul their own nest, no special burden of shame or guilt need fall upon us for the present and future condition of our world. As we discover and encounter "the wages of overshoot," mankind's humane tendencies will be strained to the breaking point. They will need the solid reinforcement they can obtain from *knowledge* that our species has not been unique in proliferating beyond carrying capacity. It will be essential to realize that what is happening to us is a mere sequel to our past achievements.

As we reap the whirlwind of troubles necessitated by excessive success, thinking ecologically of our global predicament may reduce the temptation to hate those who seem to be trespassing against us.

Notes

1. Swing 1964, p. 137.
2. "Desertification," the making of deserts in previously productive places, is a modern word for just one type of humanly relevant example of such environmental degradation. See Glantz 1977.
3. As Ophuls (1977, p. 222) says, in terms of values derived from four centuries of "progress," we now face a great deal of "retrogression." Heilbroner (1974, p. 132) foresees "convulsive" change. Ehrenfeld (1978, p. 259) believes global economic depression "without war if that is possible" is the best that can be hoped for now.
4. See Brown 1978; Ophuls 1977, p. 133; Woodwell 1970.
5. See Ophuls 1977, p. 134; Catton 1976a.
6. Cf. Ehrenfeld 1978; Watt et al. 1977, Ch. 2.
7. See Ophuls 1977. Chs. 4, 5, 6.
8. This point is developed in Ch. 10, "Industrialization: Prelude to Collapse."
9. Poll results have reflected a loss of faith in the several branches of government, disillusionment with educational and religious institutions, with the news media, and with business. The erosion of confidence ran deep enough so that President Carter alluded to it as a basic reason for American inability to unite and resolve such matters as the energy problem; see *New York Times*, July 17, 1979, p. A15. For some insights into the growth of a critical-cum-cynical outlook, see Skolnick and Currie 1976; Riesman 1976.
10. See the two articles by C. F. Wurster cited in Catton 1976b.

Selected References

Bernard, Jessie
  1973. *The Sociology of Community.* Glenview, Ill.: Scott, Foresman.
Black, Max
  1962. *Models and Metaphors.* Ithaca, N.Y.: Cornell University Press.
Brown, Lester R.
  1978. *The Twenty-ninth Day.* New York: W. W. Norton.
Catton, William R., Jr.
  1972. "Sociology in an Age of Fifth Wheels." *Social Forces* 50 (June): 436–447.
  1976a. "Toward Prevention of Obsolescence in Sociology." *Sociological Focus* 9 (January): 89–98.
  1976b. "Can Irrupting Man Remain Human?" *BioScience* 26 (Apr.): 262–267.
Ehrenfeld, David
  1978. *The Arrogance of Humanism.* New York: Oxford University Press.

Glantz, Michael H., ed.
  1977. *Desertification: Environmental Degradation in and Around Arid Lands*. Boulder: Westview Press.
Heilbroner, Robert L.
  1974. *An Inquiry into the Human Prospect*. New York: W. W. Norton.
Kuhn, Thomas S.
  1962. *The Structure of Scientific Revolutions*. Chicago: University of Chicago Press.
Lakatos, Imre, and Alan Musgrave, eds.
  1970. *Criticism and the Growth of Knowledge*. Cambridge: Cambridge University Press.
Ophuls, William
  1977. *Ecology and the Politics of Scarcity*. San Francisco: W. H. Freeman.
Riesman, David
  1976. "Some Questions about Discontinuities in American Society." Ch. 1 in Lewis A. Coser and Otto N. Larsen, eds., *The Uses of Controversy in Sociology*. New York: Free Press.
Scheffler, Israel
  1967. *Science and Subjectivity*. Indianapolis: Bobbs-Merrill.
Skolnick, Jerome H., and Elliott Currie, eds.
  1976. *Crisis in American Institutions*. 3rd ed. Boston: Little, Brown.
Swing, Raymond
  1964. *Good Evening*. New York: Harcourt, Brace & World.
Watt, Kenneth E. F., Leslie F. Malloy, C. K. Varshney, Dudley Weeks, and Soetjipto Wirosardjono
  1977. *The Unsteady State: Environmental Problems, Growth, and Culture*. Honolulu: University Press of Hawaii, for the East-West Center.
Woodwell, George M.
  1970. "The Biospheric Packing Problem." *Ecology* 51 (Winter): 1.

II | Eventually Had Already Come Yesterday

. . . any area of land will support in perpetuity only a limited number of people. An absolute limit is imposed by soil and climatic factors in so far as these are beyond human control, and a practical limit is set by the way in which the land is used.

If this practical limit of population is exceeded, without a compensating change in the system of land usage, then a cycle of degenerative changes is set in motion which must result in deterioration or destruction of the land and ultimately in hunger and reduction of the population.

> — William Allan
> *Studies in African Land Usage*
> *in Northern Rhodesia*, p. 1

The carrying capacity for human populations has been forced upward in a progressive series of steps. . . . Each step represented some cultural advance, like elimination of some competitor or over-exploitation of some resource, with a consequent overriding of the previous regulatory mechanisms restricting the population growth. . . . As the ecosystems . . . could not receive any greater input of energy, this increased . . . population could only be achieved in one of two ways. It could be effected if there were a corresponding reduction in the biomass of competing species populations. . . . Or it could be achieved by "mining" accumulated resources of the ecosystem. . . .

> — Arthur S. Boughey
> *Man and the Environment*
> (2nd ed.), pp. 251, 254

To a people who had known no other restrictions on their hunting than those imposed by the nature of their crude weapons, the thought that it might be possible [with rifles] to kill *too many* deer did not occur.

> — Farley Mowat
> *The Desperate People*, p. 20

2

The Tragic Story of Human Success

Origins of Man's Future

We are already living on an overloaded world.[1] Our future will be a product of that fact; that fact is a product of our past. Our first order of business, then, is to make clear to ourselves how we got where we are and why our present situation entails a certain kind of future.

To this purpose, consider the information about the human saga assembled in Table 1. Taken a row at a time, this table tells an enormous (and enormously revealing) story. It is the story of a world that has again and again approached the condition of being saturated with human inhabitants, only to have the limit raised by human ingenuity.

The first several rounds of limit-raising were accomplished by a series of technological breakthroughs that took almost two million years. These breakthroughs enabled human populations repeatedly to take over for human use portions of the earth's total life-supporting capacity that had previously supported other species. The most recent episode of limit-raising has had much more spectacular results, although it enlarged human carrying capacity by a fundamentally different method: the drawing down of finite reservoirs of materials that do not replace themselves within any human time frame. Thus its results *cannot be permanent*. This fact puts mankind out on a limb which the activities of modern life are busily sawing off.

In the Beginning

Some two million years ago, as represented in the first row of Table 1, creatures of another species—human, but not our kind of human— had evolved from prehuman ancestors by finding themselves more and more adapted to a place in the web of life somewhat different from the place their ancestors had occupied. They had discovered somehow that they could use (rather than merely avoid) fire; they could warm

TABLE 1.

Date	World population in millions	Most advanced economic type	Limit-raising technology	Population increase	Generations elapsed	Increase per generation
2 million B.C.		hunting and gathering	use of fire, tool-making			
					78,600	
35,000 B.C.	3[a]		spear-thrower, bow and arrow			
				167%	1,080	0.09%
8000 B.C.	8[b]	horticultural	cultivation of plants			
				975%	160	1.50%
4000 B.C.	86[c]		metallurgy (bronze)			
3000 B.C.	?	agrarian	plow	249%	160	0.78%
1000 B.C.	?		iron tools			
1 A.D.	300[d]					
				12%	55.9	0.20%
1398 A.D.	336[e]		hand firearms			
				188.4%	16.1	6.80%
1800 A.D.	969[f]	industrial	fossil fueled machinery	41.5%	2.6	14.28%
1865 A.D.	1371[g]		antiseptic surgery, etc.			
				191.8%	4.4	27.55%
1975 A.D.	4000[h]					

[a] Arthur S. Boughey, *Man and the Environment,* 2nd ed. (New York: Macmillan, 1975), p. 251.

[b] Ansley J. Coale, "The History of the Human Population," *Scientific American* 231 (Sept., 1974):43.

[c] Edward S. Deevey, Jr., "The Human Population," *Scientific American* 203 (Sept., 1960):196.

[d] Midpoint of range of estimates evaluated by John D. Durand, "The Modern Expansion

themselves with it, ward off predators with it, cook with it and thus render digestible certain organic substances that would not otherwise have been available to their bodies as nutrition. Whatever the world's capacity had been for supporting their prehuman ancestors, there was now an *additional* place for the human descendants of those earlier creatures. Their human traits enabled them to live partly upon portions of the world's substance not usable by their forebears.

These newly human beings had also begun to make and use simple tools. Moreover, they could *teach* their progeny how to make and use these artifacts. Each generation did not have to rediscover independently the techniques that had contributed to its parents' survival. Still, the accumulation of adaptive culture would have been prodigiously slow at first, and for hundreds of thousands of years there could not have been very many of these creatures. Even with fire, tools, and traditions, these humans remained what their prehuman ancestors had been: consumers of naturally available foodstuffs obtained from wild sources by hunting and gathering.

There were no census bureaus in Paleolithic times, of course. But by knowing the dependence of early man upon wild food sources, we can make reasonable estimates of maximum feasible average population density, and can estimate the extent of the earth's land area capable of supporting such hunters and gatherers. The important fact that emerges is that there could never have been very many millions of them.[2] Nevertheless, these early humans were *successful;* they survived, reproduced, adapted, and continued evolving.

By the time almost 80,000 generations of human hunters and gatherers had lived, their biological and cultural responses to the selection pressures imposed by their spreading habitats had given rise to a descendant population with essentially the inheritable physical traits we see among men and women today. Thus by about 35,000 B.C., the humans on earth were of our own species, *Homo sapiens.* Probably about three million of them were living by gathering and hunting.

---

of World Population," in Charles B. Nam, ed., *Population and Society* (Boston: Houghton, Mifflin, 1968), p. 110. See also Coale 1974, p. 43.

[e] Estimate obtained by interpolating backwards from an estimate of 350 million for 1500 A.D., using annual growth rate of .041%; see Max Petterson, "Increase of Settlement Size and Population Since the Inception of Agriculture," *Nature* 186 (June, 1960):872.

[f] Midpoint of range of estimates, Durand 1968, p. 110.

[g] Estimate obtained by interpolation, using midpoints of ranges of estimates for 1850 and 1900 given ibid.; constant exponential growth in that half-century assumed.

[h] United Nations *Demographic Yearbook* 1977, p. 115.

Increased Hunting Proficiency

We cannot really say that three million was the maximum number the
rth could ever have supported in the manner in which they were
then living. Still, we can be reasonably sure, from their slow attainment of even that number, that the earth's carrying capacity for that
kind of creature with that kind of lifestyle was not *much* greater than
that figure. However, the gradually evolving cultures of *Homo sapiens*
eventually *increased the earth's human carrying capacity.*

About 35,000 B.C., someone discovered how much harder and farther a spear could be thrown if the thrower effectively lengthened his
arm by fitting the end of the spear into a socket in the end of a hand-
held stick. Someone else invented a way of propelling miniature
spears (arrows) not only faster, but also in a manner that permitted
line-of-sight aiming, by fitting their notched ends to a cord tied to the
two ends of a springy stick. Using tools like the spear-thrower and the
bow and arrow, humans became more proficient hunters, and more of
the earth's game animals became nourishment for human bodies.

With these technological breakthroughs, the worldwide population of *Homo sapiens* increased in a little over one thousand generations from about three million souls to about eight million.[3] The total
human biomass on earth had more than doubled. Still, most of the
people in each of those thousand generations would have been utterly
unaware of increase, for, as the entry in the far right-hand column of
the table shows, each tribe was enlarged on the average by less than
1/10 of one percent during one generation—i.e., during roughly the
quarter century it took for each new parent to raise his own children
and reach grandparent status.

Learning to Manage Nature

But the time came, eventually, for *another* major breakthrough and
another enlargement of the earth's human carrying capacity. Somewhere, some of the people who gathered wild seeds for grinding into
flour observed that seeds spilled on moist earth near where the family
carried on its activities sprouted into plants that grew at least as well
as those in the wild. In time these plants would bear a new crop of
seeds, conveniently harvestable. *Homo sapiens* went on to develop
this discovery into techniques of plant cultivation, effecting a major
transformation of the relation of our species to nature's web of life.

Henceforth, some of us were going to obtain nourishment from a *humanly managed portion* of the biotic community, rather than merely gathering the products of plant and animal species that we could use if we reached them before other consumer animals or invisible decomposer organisms.

This horticultural revolution, by which hunters and gatherers turned into farmers, was followed by a tenfold increase in the earth's human population.[4] This increase occurred in 1/6 as many generations as the previous increase phase. Such acceleration indicates that mankind's daring to undertake the management of a portion of nature had again raised the earth's human carrying capacity. Biologically, this species, with the remarkable capability of achieving cultural innovations, was proving a resounding success.

It began to be possible for a minuscule but increasing fraction of any human tribe to devote its time to activities other than obtaining sustenance. Human social organization could begin developing along more elaborate lines, and the rate of cultural innovation could further accelerate. Each increment of technology gave mankind a competitive edge in interspecific competition. Our species was well on its way to being the dominant member of the ecosystem.

Compound Interest

Note that, even after this horticultural acceleration of population growth, change would have remained almost unnoticeable to those living through it. The increment in an average generation was still a mere 1.5 percent. The starting population of 8 million was, in effect, multiplied in one generation by a factor of 1.015, and then that product was again multiplied in the next generation by 1.015, and so was that product, and so on. The "interest" of 1.5 percent on the initial "investment" was compounded by each generation—160 times between 8000 B.C. and 4000 B.C. Thus:

$$8,000,000 \times (1 + 0.015)^{160} = 86,000,000, \text{ approximately.}$$

So the numbers shown in the "generations elapsed" column of Table 1 are more than just expressions of the time intervals between the dates shown in the first column; they must be read as exponents applied to multipliers that are derived from the figures in the last column. Even at low percentage rates of increase per generation, the "compound interest" pattern can produce great change when enough generations elapse.

21

As advancing human culture extended the niches available to mankind, recurrent surges of essentially exponential growth in numbers became possible. (The well-known "population explosion" of our own time was merely the most recent episode in a process that has been going on since antiquity.)

Tools, Organization, and Standard of Living

By about 4000 B.C., stone and bone tools began to be augmented and then superseded by metal tools as *Homo sapiens* moved into what his history-writing descendants would one day label the Bronze Age. This enhancement of man's tool kit was followed by further population increase. Metallurgy enhanced the ability of the human species to harvest nature's products, rather than leaving them to be used by other consumer species. It also gave further impetus to the elaboration of a "division of labor" among increasingly specialized occupations. From here on, the growth of organization among humans would be an increasingly important factor in their dominance over the environment supporting them.[5]

If cultural innovations were to cease, or if some ultimate limit proved impossible to transcend by cultural progress, exponential growth would give way to a curve of diminishing returns. Limited carrying capacity would reduce the rate of growth in successive generations. Eventually, as population approached carrying capacity, the growth rate would approach zero—of necessity. That is what "carrying *capacity*" means.[6]

But innovations continued, and the ceiling *was* raised again. Around 3000 B.C., man the cultivator of plants went in for an early version of "mass production," tilling land in larger tracts than before. This was made possible by invention of the plow, which enabled the farmer to begin using non-human energy to turn over the soil—energy supplied by the muscles of an ox or a horse, though at first a plow was sometimes pulled by a slave or a wife (and had to be rather small). One farmer could manage more soil with this additional tool. But an agriculture that used draft animals had to use some of its land to raise crops to be eaten by those animals, so this new technology would not immediately raise human carrying capacity as dramatically as previous innovations had done.

There was also an alternative use for this particular increment in sustenance-producing power. A farmer with a plow and a draft animal could farm enough land to feed himself, the animal, his own family, and perhaps have a bit to spare. So some small but gradually increasing fraction of the population could now do things other than raise food. Human groups could opt for further elaboration of their lives, rather than for simple expansion of their numbers.

About 1000 B.C., iron tools began to supplement and replace those made of bronze. Again, some of the carrying capacity increment was used to enhance, little by little, the standard of living of at least some groups.

The separate effects of these last several innovations upon population increase cannot be assessed, because usable estimates of population numbers at the times these new tools and techniques came into use are not available.[7] But between the beginning of the Bronze Age and the birth of Christ (a date for which there does happen to be a more or less agreed upon population estimate) their cumulative effect was to expand the world's human stock from about 86 million to about 300 million—an average rate of increase of about ¾ of one percent per generation. Slower increase continued for another millennium.

Firearms

Then came a different kind of breakthrough. Early in the fourteenth century firearms were invented, and were immediately put to military use. The first firearms were hardly portable, and hardly suitable for any non-military purpose. If they were to have any effect on carrying capacity, that effect had to be indirect. By changing the nature of warfare they would eventually change the nature of political organization, which would, in turn, alter the way human populations would relate themselves to the resources of the world around them.

Within three generations after these first firearms came into use, hand-carried firearms began to be made. Since these could have had some direct bearing upon human ability to harvest meat, they (rather than their more cumbersome military forerunners) are given a place in Table 1. In the next sixteen generations, we see a higher average rate of population increase than ever before. It is too high, in fact, to be solely due to improved game-harvesting efficiency. It came about quite differently.

The cumulative effects of human increase over the past two million years were becoming significant. The portions of the earth's land surface available to those human tribes that had thus far experienced *all* of these technological breakthroughs were coming to be rather fully occupied by humans. But the tools and the knowledge available to these culturally most advanced segments of *Homo sapiens* were enabling (and causing) some men to leave the land and venture more and more daringly onto the sea. Less than a century after the invention of portable firearms, Europeans would discover lands they had not previously known existed. In the generations after that discovery, the Europeans' superiority in weapons would enable them to take possession of whole new continents whose prior human inhabitants were much less numerous, because they were still living mostly at the Stone Age hunter-gatherer or early horticultural level.

Firearms did not enlarge the planet. However, they served to enlarge once again the carrying capacity of the world known to Europeans, by making available for settlement and exploitation a "virgin" hemisphere. The expansion of territory available for use by Europe's already advanced means is the main reason why firearms can be said to have led to the unprecedented rate of increase in human numbers during this last portion of the agrarian period.

Abundance

I shall call the centuries that followed the sudden expansion of European man's habitat by voyages of discovery the *Age of Exuberance*, for reasons to be spelled out in later chapters. During that age, man largely forgot that the world (i.e., Europe) had once been saturated with population, and that life had been difficult for that reason. Discovery of the New World gave European man a markedly changed relationship to the resource base for civilized life. When Columbus set sail, there were roughly 24 acres of Europe per European. Life was a struggle to make the most of insufficient and unreliable resources. After Columbus stumbled upon the lands of an unsuspected hemisphere, and after monarchs and entrepreneurs began to make those lands available for European settlement and exploitation, a total of 120 acres of land per person was available in the expanded European habitat—five times the pre-Columbian figure![8]

Changelessness had always been the premise of Old World social systems. This sudden and impressive surplus of carrying capacity shattered that premise. In a habitat that now seemed limitless, life could be lived abundantly. The new premise of limitlessness spawned new beliefs, new human relationships, and new behavior. Learning was advanced, and a growing fraction of the population became literate. There was a sufficient per capita increment of leisure to permit more exercise of ingenuity than ever before. Technology progressed, and technological advancement came to be the common meaning of the word "progress."

But the aura of limitless opportunity had another effect: further acceleration of population growth. To go into some details not shown explicitly in Table 1, between 1650 and 1850, a mere two centuries, the world's human population doubled. There had never before been such a huge increase in so short a time. It doubled again by 1930, in only eighty years.[9] And the next doubling was to take only about forty-five years! As people and their resource-using implements became more numerous, the gap between carrying capacity and the resource-use load was inevitably closed. American land per American citizen shrank to a mere 11 acres—less than half the space available in Europe for each European just prior to Columbus's revolutionizing voyage. Meanwhile, per capita resource appetites had grown tremendously. The Age of Exuberance was necessarily temporary; it undermined its own foundations.

Most of the people who were fortunate enough to live in that age misconstrued their good fortune. Characteristics of their world and their lives, due to a "limitlessness" that had to be of limited duration, were imagined to be permanent. The people of the Age of Exuberance looked back on the dismal lives of their forebears and pitied them for their "unrealistic" notions about the world, themselves, and the way human beings were meant to live. Instead of recognizing that reality itself had actually changed—and would eventually change again—they congratulated themselves for outgrowing the "superstitions" of ancestors who had seen a different world so differently. While they rejected the old premise of changelessness, they failed to see that their own belief in the permanence of limitlessness was also an *over*belief, a superstition.

As the gap closed, conditions of life did change—of necessity. The world reentered an age of population pressure. Its characteristics had to resemble, in certain ways, the basic features of the Old World of pre-Columbian times. Except that now there were ever so many more human beings, all parts of the planet were in touch with each other, per capita impact on the biosphere had become enormously amplified by technology, depletion of many of the earth's non-renewable resources was already far advanced—and the inhabitants of this post-exuberant world had acquired from the Age of Exuberance expectations of a perpetually expansive life.

The Takeover Method

The Europeans who began taking over the New World in the sixteenth and seventeenth centuries were not ecologists. Although they soon were compelled to realize that the Americas were not quite *un*-inhabited, they were not prepared to recognize that these new lands really were, in an ecological sense, much more than "sparsely" inhabited. This second hemisphere was, in fact, essentially "full." As we have seen, the world supported fewer people when they were at the hunter-gatherer level than when they advanced to the agrarian level. In the same way, a continent that was (ecologically speaking) "full" of hunters and gatherers was bound to seem almost empty to invaders coming from an agrarian culture and accustomed to that culture's greater density of settlement.

Ethnocentrism prevented most Europeans from seeing themselves as they must have appeared to the Indians—as competitors for resources the Indians were already exploiting as fully as they knew how. Ecologically, these vast "new" lands did not have "plenty of room" for Indians plus Europeans, as the Europeans easily supposed. Indians living by hunting-gathering and by simple horticulture were going to be *displaced* by incoming hordes of Europeans practicing advanced agrarian life.

Even if there had been less ethnocentrism, and if principles of Christian compassion had sufficed to preclude all suspicion, hostility, and bloodshed in the interactions between "civilized" and "savage" peoples, total ignorance of the ecological implications of different levels of technology would have enabled the takeover to occur. Europeans were able to move to the New World with no pangs of conscience about relegating the native peoples to a shrinking fraction of these continents. The shrinking fraction afforded insufficient carrying capacity (when exploited by hunting and gathering or by primitive horticulture) to accommodate the number of Indians already generated by their previously more extensive environment. But neither the concept of carrying capacity nor its relation to stages of human culture was part of the European settlers' mental equipment. So the displacement occurred.

Essentially the same displacement followed from the same ethnocentrism and ecological naivete when settlers from Europe invaded Australia and New Zealand. An approximation of this pattern also prevailed for a while as Europeans later took over the more or less temperate parts of Africa, although there a difference in the invader/native ratio eventually began to reverse the relationship with more numerous Africans eventually beginning to oust Europeans.

All over the world, Europeans had acted on the premise that it was only fair and reasonable for "unused" or "underused" lands (i.e., lands being used by non-agrarian non-Europeans) to be "put to good use." In the absence of ecological understanding, that premise had seemed utterly sound.

The takeover method of enlarging carrying capacity was far older than the Age of Exploration and the centuries of colonial expansion. Invading and usurping lands already occupied by others was essentially what mankind had been doing ever since first becoming human. Each enlargement of carrying capacity reviewed in the preceding pages consisted essentially of *diverting* some fraction of the earth's life-supporting capacity from supporting other kinds of life to supporting our kind. Our pre-*sapiens* ancestors, with their simple stone tools and fire, took over for human use organic materials that would otherwise have been consumed by insects, carnivores, or bacteria. From about 10,000 years ago, our earliest horticulturalist ancestors began taking over *land* upon which to grow crops for human consumption. That land would otherwise have supported trees, shrubs, or wild grasses, and all the animals dependent thereon—but fewer humans. As the expanding generations replaced each other, *Homo sapiens* took over more and more of the surface of this planet, essentially at the expense of its other inhabitants. At first those displaced were creatures with teeth and claws instead of tools, with scales or feathers or fur instead of clothes.

In this takeover process, man was behaving as all creatures do.[10] Each living species has won for itself a place in the web of life by adapting more effectively than some alternative form to a given role. What is true of a species is also true of a subdivision within a species. A given tract of land has greater carrying capacity for the subspecies that can extract more from it than for other portions of the species that happen to be less equipped to exploit it.

None of this is said for the sake of *justifying* displacement of American Indians (or Polynesians, Aborigines, or Africans) by Europeans. Recently aroused pangs of guilt have made European-descended Americans more conscious of the suffering of those who were displaced. Although guilt feelings cannot resurrect the Indians who were forced to yield their place to more powerfully equipped Europeans, perhaps such feelings can prompt us to think about matters we might otherwise have continued to neglect. By *explaining* this human displacement episode as a special case of the ecological principle of "competitive exclusion," we can at least take note of how common the takeover process has been in the ecological history of the world. Then, having seen that, we should also be able to see how fundamentally different the takeover method was from another method by which human carrying capacity has been most recently stretched. Recognition of the difference is essential to understanding the human predicament.

The Drawdown Method

About 1800 A.D., a new phase in the ecological history of humanity began. Carrying capacity was tremendously (but temporarily) augmented by a quite different method; takeover gave way to drawdown. A conspicuous and unprecedentedly large acceleration of human population increase got under way as *Homo sapiens* began to supersede agrarian living with industrial living.

Industrialization made use of fossil energy. Machinery powered by the combustion of coal, and later oil, enabled man to do things on a scale never before possible. New, large, elaborate tools could now be made, some of which enhanced the effectiveness of the farming that of course had to continue. Products of farm and factory could be transported in larger quantity for greater distances. Eventually the tapping of this "new" energy source resulted in the massive application of chemical fertilizers to agricultural lands. Yields per acre increased, and in time acreages applied to the growing of food for humans were substantially increased—first by eliminating draft animals and their requirements for pasture land, but also by reclaiming land through irrigation, etc.

This time mankind was not merely taking away from competitors an additional portion of the earth's life-supporting capacity. (He was still doing this, and still not recognizing that this was what he had always done. But—worse—he was now also not recognizing the true nature of something else he was doing on a vast scale. So man was painting himself into a corner.) This time, the human carrying capacity of the planet was being supplemented by digging up energy that had been stored underground millions of years ago, captured from sunlight which fell upon the earth's green plants long before this world had supported any mammals, let alone humans, or even prehuman primates. The solar energy had been captured by photosynthesis in plants that grew and died and were buried during the Carboniferous period, without the efforts of any farmers. (As we shall see in the next chapter, the fact that no farm labor had to be paid to raise the Carboniferous vegetation, and that no investments in farm machinery used to grow those prehistoric "crops" had to be amortized, etc., helped get us into our present predicament.)

Carrying capacity was this time being augmented by drawing down a finite reservoir of the remains of prehistoric organisms. This was therefore going to result in a *temporary* extension of carrying capacity; in contrast, previous enlargements had been essentially permanent, as well as cumulative.

Being impermanent, this rise in apparent carrying capacity begged one enormously important question: What happens if population, as usual, increases until it nearly fills this temporarily expanded set of opportunities, and then, because the expansion was only temporary, the world finds itself (like the Indians on their shrunken territories) with a population excess? What are the implications of a carrying capacity deficit for mankind's future? What happens, for example, when supplies of oil become scarce, when tractor fuel becomes unavailable or prohibitively expensive, and when farmers again have to take ¼ to ⅓ of the land on which they now raise food for humans and convert it instead to raising feed for draft animals?[11]

Such questions were not asked as long as we viewed our world with a pre-ecological paradigm. The myth of limitlessness dominated people's minds. Had anyone conceived such implausible-seeming questions in the Age of Exuberance, the answer might have seemed equally incredible: post-exuberant nations and individuals would have a compulsive need to deny the facts so as to deny their own redundancy. (We shall examine such denial of the new reality in Part III of this book, and again in Part V.)

Industrialization came about at a fast enough pace so that it enlarged per capita wealth and was not entirely devoted to enlarging population. In principle, any increase in carrying capacity—temporary or permanent—affords a choice between enabling the same number of individuals to live more lavishly or enabling a larger number of individuals to live at previous standards. When the enlargement of carrying capacity is modest and is spread over many generations, it tends to be used mainly to increase numbers; if it is enormous and comes so suddenly that human numbers just don't rise at the same pace, it raises living standards. The European takeover of the New World had enlarged carrying capacity (for Europeans) just fast enough to begin having this salutary effect. By drawing down stores of exhaustible resources at an ever-quickening pace, industrialization (temporarily) augmented carrying capacity even faster, affording opportunity for quite a marked rise in prosperity *and* for a phenomenal acceleration of population increase. The welcome rise in prosperity reinforced the dangerous myth of limitlessness and obscured for a while the hazards inherent in the population increase.

Overshoot Aggravated

Scarcely more than two generations had tasted the fruits of industrialization when the growth of population was still further accelerated by truly effective death control. The role of micro-organisms in producing diseases was discovered. In 1865 the practice of antiseptic surgery began. It serves in Table 1 as a reasonable demarcation of the beginning of an era filled with related breakthroughs in medical technology: hygienic practices, vaccination, antibiotics, etc. The total effect of this recent series of achievements has been to emancipate mankind more and more from the life-curtailing effects of the invisible little creatures for which human tissues used to serve as sustenance. Like other prey species newly protected from their predators, we have been fruitful and have so multiplied that we have much more than "replenished" the earth with our kind.

These achievements in death control re-channeled the effects of industrialization; they increased the rate at which human population could increase. More of the unprecedentedly rapid rise in apparent carrying capacity resulting from industrial drawing down of resource stocks was devoted to supporting population growth, and less was devoted to supporting enhanced living standards, than might otherwise have been the case.

Death control was a real boon to the first three or four generations that experienced it. Increasingly, parents were spared bereavement during their child-rearing years, and people of all ages were spared the suffering and debilitation that infectious diseases used to inflict. Fewer children became orphaned. Fewer adults became widowed in the prime years of life.

But all these benefits helped us to overshoot *permanent* carrying capacity.[12] For most people, as this was happening, "carrying capacity" remained an unknown phrase. The concept was absent from the paradigm by which people in the Age of Exuberance perceived and understood their world. Industrialism had given us a temporary increase in opportunities—a very dangerous blessing. Death control gave us a further rapid increase in population not based on a further rise in carrying capacity. Thus, in the seven generations since 1800, world population quadrupled, and mankind came into a really precarious situation.

The precariousness remained unseen by many. Looking back on a century or two of remarkable technical achievements, accompanied by growth of human numbers that was itself culturally defined as a kind of progress (as every town aspired to become a city), minds that had not yet learned the distinction between methods of boosting carrying capacity and methods of overshooting it foresaw no insurmountable difficulty in simply repeating past breakthroughs. It was imagined, for example, that "fast breeder reactors" and other technological eggs-not-yet-hatched could be counted on to provide further increments of carrying capacity whenever nature's limits began to hurt. (This attitude will be given a suggestive name in Chapter 4 and explored further in Chapter 11.)

During World War II, the brashly American words of a popular song proclaimed: "We did it before, and we can do it again!" A generation after that conflict, we seemed to be taking a demilitarized version of that cliché as the basis for presupposing the supportability of further increases in the population-technology load upon finite environments. People displayed either persistent ignorance of the carrying capacity concept, or naive faith that carrying capacity could always be expanded, that limits could always be transcended. Such an assumption seemed to underlie the stubborn refusal of capitalists and Marxists alike to acknowledge that the myth of limitlessness had at last become obsolete. There was also the assumption that further advances in technology would necessarily enlarge carrying capacity, not reduce it. Enlargement of carrying capacity had been the role of technology in the past; however, we shall see (in Chapter 9) that there has been a reversal of this role in the industrial era. Technology has enlarged human appetites for natural resources, thus diminishing the number of us that a given environment can support.

Back to Hunting and Gathering

The breakthrough we call industrialism was fundamentally unlike earlier ones. It did not just take over for human use another portion of the web that had previously supported other forms of life. Instead, it went underground to extract carrying capacity supplements from a finite and depletable fund—a fund that was created and buried by nature, scores of millions of years before man came along. The drawdown method that we call industrialism relied for its increase of opportunities upon use of resources that are not renewed in an annual cycle of organic growth. To expect to "do it again" is to expect to *find* other exhaustible resources each time we use up a batch of them. Only once could the technologically most advanced nations of mankind discover a second hemisphere to relieve the pressure in a filled-up first hemisphere; nevertheless, modern industrial societies have continued to behave as if massive "exploration" efforts could forever continue to "discover" additional deposits of mineral materials and fossil fuels. In short, industrial life depends on a perpetual *hunt* for required substances. To take one example, in order to continue present rates of use of copper, the United States must each year *find* 250 million tons of ore (containing 0.8 percent copper)—more than a ton for each of us.

The mineral and fuel deposits upon which we are now so dependent were put into the earth by geological processes that happen only at a pace enormously slow by human standards. Since 8000 B.C. mankind has been taking over management of contemporary botanical processes, the source of sustenance materials that have renewal times much shorter than a human lifespan. Now we rely, as members of industrial societies, upon other substances with renewal times that may be thousands or even millions of times longer than a human lifespan. Their renewal is by geological processes; present stocks of them were put in place by operation of those processes over immensely long stretches of earth history. Mankind cannot realistically hope to assume management of prehistoric events, or to replenish the ores and fuels now being extracted so ravenously. Instead, we must face the fact that, after ten millennia of progress, *Homo sapiens* is "back at square one." Industrialization committed us to living again, massively, *as hunters and gatherers* of substances which only nature can provide, and which occur only in limited quantity.

A major oil company whose credit card has been a convenience to me in my travels has recently confirmed this—unwittingly, of course—by printing at the bottom of my monthly statement a bit of institutional advertising. In an effort to enlist customer support for its resistance to congressional pressures against combined ownership of both "production" and "marketing" facilities, this company's message proclaims that it "does the whole job—*finding and delivering* oil products you need" (my italics).

Our species had been an enormous biological success. But success carried to excess can be disastrous. The shift from takeover to drawdown actually yielded excessive success. As we shall see, this situation has had a natural sequel. Much of the turmoil so vexing to the generation that saw the fourth billion added to the world's human population can be understood in such terms. We had already begun to encounter the penalties of becoming again what our remote ancestors were—consumers of substances provided by nature and not by man, substances we obtain from sources not subject to replenishment by our manipulations. We became heavily dependent upon hunting for natural deposits of these substances, and upon continually gathering vast quantities for our use. Euphemistically calling the new versions of these ancient activities "finding" and "delivering," or "exploration" and "production," only blinded us to what we were doing. It did not protect us from the consequences.

Notes

1. Calculations supporting this statement appear in the next chapter; the present chapter tells the story of our arrival in this predicament. For another statement indicating that this is indeed the nature of our situation, see Kingsley Davis, "Zero Population Growth: The Goal and the Means," *Daedalus* 102 (Fall, 1973):26.
2. For examples of the reasoning behind any inference as to the size of prehistoric populations, see Hollingsworth 1969; Ehrlich, Holm, and Brown 1976, p. 457; Coale 1974, p. 41; Desmond 1962, pp. 3–4.
3. For documentation of this and subsequently mentioned population estimates, see sources cited in the notes for Table 1.
4. Estimates of prehistoric world populations are less exact than modern population figures, of course; but the increase discussed in this paragraph would be no less significant if its magnitude were appreciably less or somewhat more than stated.
5. See Childe 1951, pp. 25–26.
6. In technical terms, carrying capacity is represented by the upper limit of an S-shaped logistic growth curve, into which an initially exponential growth curve gets converted by the finiteness of the habitat and its resources.
7. It was not until the latter part of the seventeenth century that scientific study of population began. A British mathematician, John Graunt, in 1662 studied parish clerks' records of baptisms and burials, and derived sex ratios, fertility ratios, measures of natural increase, etc. In 1693 the astronomer Edmund Halley constructed a life-expectancy table from church records.
8. Webb 1952, pp. 17–18.
9. Desmond 1962, p. 12.
10. See the brief comments in Boughey 1975, p. 17, on "competitive exclu-

sion" and "resource partitioning," and the more extensive exposition by Hardin 1960. To recognize the displacement of one population of humans by another (with more advanced technology) as an instance of this common ecological process, it is useful to think in terms of a concept developed in Chs. 6 and 9, "quasi-speciation."

11. From 1973, as shortages of fossil energy came to public attention, it was often supposed that "energy plantations" would afford a solution. The fact that this would put fuel-burning engines into the same competitive relation with food-consuming humans that formerly applied to farmers' draft animals was almost universally overlooked.

12. Since "carrying capacity" is by definition the maximum *permanently* supportable population, the expression "permanent carrying capacity" is redundant. The redundancy may serve, nevertheless, to underscore the nature of our predicament. A related point is made by introducing in the next chapter the concept of "phantom carrying capacity" to refer to such things as fossil energy; to speak of "temporary carrying capacity" would be a contradiction.

Selected References

Ackerknecht, Erwin H.
1968. *A Short History of Medicine*. Rev. ed. New York: Ronald Press.
Borrie, W.D.
1970. *The Growth and Control of World Population*. London: Weidenfeld and Nicholson.
Boughey, Arthur S.
1975. *Man and the Environment*. 2nd ed. New York: Macmillan.
Childe, V. Gordon
1951. *Social Evolution*. New York: Henry Schuman.
1954. *What Happened in History*. Rev. ed. Harmondsworth, Middlesex: Penguin Books.
Coale, Ansley J.
1974. "The History of the Human Population." *Scientific American* 231 (Sept.):41–51.
Deevey, Edward S., Jr.
1960. "The Human Population." *Scientific American* 203 (Sept.):194–204.
Desmond, Annabelle
1962. "How Many People Have Ever Lived on Earth?" *Population Bulletin* 18 (Feb.):1–19.
Ehrlich, Paul R., Richard W. Holm, and Irene L. Brown
1976. *Biology and Society*. New York: McGraw-Hill.
Hardin, Garrett
1960. "The Competitive Exclusion Principle." *Science* 131 (Apr. 29):1292–97.

Hollingsworth, T. H.
　1969. *Historical Demography*. Ithaca, N.Y.: Cornell University Press.
Lenski, Gerhard, and Jean Lenski
　1978. *Human Societies: An Introduction to Macrosociology*. 3rd ed. New York: McGraw-Hill.
Mumford, Lewis
　1934. *Technics and Civilization*. New York: Harcourt, Brace.
Nam, Charles B., ed.
　1968. *Population and Society*. Boston: Houghton Mifflin.
Potter, David
　1954. *People of Plenty: Economic Abundance and the American Character*. Chicago: University of Chicago Press.
Singer, Charles, E. J. Holmyard, and A. R. Hall, eds.
　1954. *A History of Technology*. 5 vols. Oxford: Clarendon Press.
Ubbelohde, A. R.
　1955. *Man and Energy*. New York: George Braziller.
Webb, Walter Prescott
　1952. *The Great Frontier*. Boston: Houghton Mifflin.

3 | Dependence on Phantom Carrying Capacity

Other Foundations, Other Limits

Because the people of industrial nations did not recognize themselves as hunters and gatherers, they adhered to premises that were becoming more and more false. Franklin D. Roosevelt spoke for all believers in those premises in the next-to-last sentence he ever wrote: "The only limits to our realization of tomorrow will be our doubts of today."

Six years before Roosevelt's final expression of the optimistic faith that had become standard in the Age of Exuberance, one of the world's foremost demographers, P. K. Whelpton, had written that increasing numbers of people were only compatible with a rising standard of living when a nation either was still underpopulated, or could still call upon technological progress to offset the disadvantages of overpopulation. According to Whelpton, the United States in 1939 was *already overpopulated*.[1] Technology, which had formerly enlarged carrying capacity, was growing in its power to do just the opposite—to increase per capita resource requirements, and thus aggravate the overload.

Still, assumptions and expectations from the Age of Exuberance persisted for another generation, making more convulsive than it might have been the eventual change entailed by their obsolescence.[2]

Roosevelt died in 1945 without recognizing the end of exuberance. He was drafting a Jefferson Day radio address when a cerebral hemorrhage struck him down. His final sentence was: "Let us move forward with strong and active faith." Under his leadership, actions by the American nation (in concert with many others) had done much to renew the commitment of people to this exuberant spirit, delaying for another generation widespread comprehension of its obsolescence. Strong and active faith was characteristic of the age of apparent limitlessness; it had motivated nation-building and other impressively creative human activities. After World War II, for one more generation, people in many parts of the world would act from the illusion that the world's less fortunate could reap the benefits of an age of neo-exuberance by creating new nations in areas formerly held as colonies by one European power or another. But terminating colonialism could not renew limitlessness. Both imperialism and the subsequent graduation of the earliest and richest colonial components of empires into the status of new nations had been results, not causes, of the age of surplus carrying capacity.

The achievements of *Homo sapiens* have always required foundations other than the self-assurance and determination to which Roosevelt appealed. Sheer will-power, important as it can be, cannot be implemented without material resources and physical energy, regardless of the institutional expectations of a people. As long ago as 1893, at a meeting of the American Historical Association, Frederick Jackson Turner insisted that "Behind institutions, behind constitutional forms and modifications, lie the vital forces that call these organs into life and shape them to meet changing conditions."[3] In the special instance of American institutions, these vital forces had consisted very largely, Turner said, of the presence of free land and the continuous westward advancement of American settlement by European immigrants and their descendants into areas previously inhabited only by disregardable non-Europeans.

When the land was filled up and no longer available at little or no cost, and when its inhabitants were no longer people who could be disregarded, institutions had to change. But persistent myths would delay institutional adaptations that were eventually inevitable. These obsolete myths would impede understanding of the real causes of change.[4]

No Longer Hypothetical

Here was a characteristic instance of cultural lag: by the time a substantial number of people began to worry out loud about what to do "if" the world "eventually" were to become overloaded, it already was. In the 1960s many books and articles appeared which spoke of dire troubles ahead "unless" growth of population were halted, or "if" the rate of extraction of petroleum or other resources from the earth continued to double every N years. Some of this kind of literature had come out in the 1950s, and one emphatic treatise on the subject— William Vogt's *Road to Survival*—was published in 1948.[5] The implication in nearly every one of these publications was that the dire troubles were still hypothetical, a possibility still avoidable, provided the right corrective measures were adopted in time. The purpose of most writers was to arouse people to accept or demand the necessary preventives before it was too late. Although some authors insisted that it was already later than people generally realized, few ventured to suggest (even in the 1970s) that, for the post-exuberant world, eventually had already come yesterday.

The growth and progress upon which we looked back with such pride had *committed* mankind to living on a scale that exceeds the sustainable carrying capacity of this finite planet, and the leaders of nations continued to devote far more effort toward attempting to prolong overshoot than toward undoing it. Reluctance to face facts was driving us to make bad matters worse. The faster the present generation draws down the fossil energy legacy upon which persistently exuberant lifestyles now depend, the less opportunity posterity will have to live in anything like the same way or the same numbers. Yet most contemporary political proposals for solving problems of economic stagnation or inequity amount to plans for speeding up the rate of drawdown of non-renewable resources.[6]

Invisible Acreage

The truth of these statements is implicit in the concept of "ghost acreage." Georg Borgstrom, a food scientist at Michigan State University, devoted a whole chapter of his 1965 book, *The Hungry Planet,* to this subject. A number of nations have seemed to get away with exceeding the human carrying capacity of their own land, but Borgstrom pointed out that they had only been able to do so by drawing upon carrying capacity that was "invisible"—i.e., located elsewhere on the planet. The food required by such a nation's population comes only partly from the harvest of "visible acreage"—farm and pasture land within the nation's borders. A very substantial fraction comes from net imports of food. Not all the imports come from other countries; some are obtained from the sea. Borgstrom therefore subdivided "ghost acreage" into two components, "trade acreage" and "fish acreage." By each phrase he simply expressed, in terms of land area, the additional farming that would have been needed to provide from internal sources the net portion of a nation's sustenance actually derived from sources outside its boundaries and in excess of its own carrying capacity. As we shall see, a third component must be recognized if we are to understand fully the part played by ghost acreage in the life of modern man.

To see the importance of Borgstrom's two components, trade and fish acreage, let us consider two examples: Great Britain, a national ancestor of the United States, and Japan, a booming industrial giant in the Far East. By 1965 more than half of Britain's sustenance was coming from ghost acreage.[7] If food could not be obtained from the sea (6.5%) or from other nations (48%), more than half of Britain would have faced starvation, or all British people would have been less than half nourished. Likewise, if Japan could not have drawn upon fisheries all around the globe and upon trade with other nations, two-thirds of her people would have been starving, or every Japanese citizen would have been two-thirds undernourished[8] (which presumably means that nearly all might have died). Yet this was the most prosperous nation in the Orient, the one whose low birth rate supposedly exemplified Asia's hope of *averting* overpopulation.

These densely populated nations had continued to exist and prosper only because, on top of their own intensive agriculture, they could harvest the oceans and could export non-agricultural products in exchange for food from countries with agricultural surpluses. Accordingly, ghost-acreage-dependent countries like these were vulnerable to foreign efforts to manipulate their policies (e.g., the Arab oil embargo). They were also threatened by population growth in the food-exporting countries, for such growth would stem the flow of food exports they needed in order to survive.

When there ceased to be agricultural surpluses anywhere, and when *all* nations became dependent on oceanic ghost acreage, population densities of a British or Japanese magnitude would be more obviously non-viable. In the meantime, Americans, Canadians, Australians, etc., habitually pointed to their own wheat surpluses and reassured themselves that *they* were a long way from being overpopulated. "Look at Japan," said their people, blinded still by the old pre-ecological paradigm: "—much more heavily populated than we are, yet prospering."

Space Age accomplishments at last brought some recognition that the earth must be considered as a unit. It is man's *one* habitat. This planet is an island, more absolutely than Japan or Britain. When *Homo sapiens* in the 1960s became able to "export" a few manufactured items from earth to the moon, to Mars, to Venus, etc., only new knowledge came back in exchange—there were no imports of foodstuffs. The knowledge increments were magnificent achievements, well worth pursuing; still, the terms of the exchange by which they were accomplished began to underscore the fact that mankind as a whole could not disregard overpopulation, as some component countries had continued to do when they outgrew the carrying capacities

39

of their own territories. There was no "trade acreage" in outer space.

"Fish acreage," if considered globally, could also be seen to provide only a shrinking reserve for the world's family of nations to fall back on. The earth's oceans are finite. In the 1970s the fish, whales, and other edible marine creatures were already being harvested in greater quantities than would permit a sustained yield.[9] From overfishing and from pollution, the seas were dying.[10] Accordingly, various nations were becoming more overtly competitive in their use of this reserve. Some were compelled by circumstances to express such competitiveness in the form of territoriality. Human societies thus turned out to behave much in the manner of communal groups of other mammal species, when one group begins to suffer from encroachments by others upon resources it needs in order to sustain itself. A typical animal response to population pressure is to assert territorial claims and to exclude competitors from the claimed area.[11] A number of nations unilaterally extended their claims to exclusive fishing. The original "three-mile limit" of national sovereignty over the seas became a "twelve-mile limit," and then various nations went on to extend their fishing claims out to fifty miles, or a hundred, or two hundred.[12]

The so-called Cod War between Britain and Iceland, and similar friction between the United States and Peru, were territorialist responses to the end of exuberance. These territorialist responses were becoming so universal that they compelled the United Nations to begin rewriting the law of the sea to institutionalize such marine claim-staking. Meanwhile, the United States unilaterally proclaimed a 200-mile fishing limit effective March 1, 1977, and this severely pinched fish acreage–dependent Japan. On a November day in 1976, when talks began that were intended to lead to a bilateral North Pacific fisheries agreement between Japan and the U.S., thousands of banner-carrying Japanese took to the streets of Tokyo in protest. In a newspaper ad, the Japan Fisheries Association said the 200-mile limit off American shores could seriously restrict Japanese protein consumption, curtailing by as much as 44 percent the amount of fish that would be eaten in Japan.[13]

Importing from the Past

The onetime American shibboleth, "Freedom of the seas," had been an idea born in the Age of Exuberance. Post-exuberant overload was now depleting the world's resources and requiring even the United States to take such steps as fencing off a private fishing domain. But the predicament was global. Without knowing it, *Homo sapiens* faced a plight much like that of Japan when confronted with fish-depleted oceans. As an island in space, the world could not rely on imports from else*where;* nevertheless, it was already heavily dependent upon imports from else*when.* That we were importing from the past becomes clear when we logically extend Borgstrom's ghost acreage concept to include a third component. Technological progress had made mankind heavily dependent upon imports of energy from prehistoric sources. Man's use of fossil fuels has been another instance of reliance on phantom carrying capacity.

The energy we obtain from coal, petroleum, and natural gas can be expressed as "fossil acreage"—the number of additional acres of farmland that would have been needed to grow organic fuels with equivalent energy content. Mankind originally did rely on organic fuels, chiefly wood. Wood was a renewable resource, though even in the world's once vast forests it grew in limited quantity. Access to vast but non-renewable deposits of coal and petroleum came to be mistaken by peoples and nations as an opportunity for permanently transcending limits set by the finite supplies of organic fuel.

When fossil fuels had been depleted enough to make supplies of them precarious, insufficient, and increasingly expensive, proposals for making up the shortfall included various versions of "energy farming"—growing crops from which fuels could be derived. The acreage required for future energy plantations is an obvious measure of the phantom carrying capacity upon which fossil-fueled civilization had been depending. As we shall see in the next section of this chapter, re-expressing modern rates of energy use in ghost acreage terms enables us to recognize how seriously our hunting-and-gathering industrial civilization overshot the real (i.e., permanent) carrying capacity of the planet's visible acreage.

Everything human beings do requires energy. At the barest minimum, animals human in form but with no technology would have been converting in their own bodies some 2,000 to 3,000 kilocalories of chemical energy (from food) into heat in the course of a day's activity. With the mastery of fire and with the domestication of animals, additional energy came under human use, even before some of the energy from flowing water and moving air began to be harnessed for the conduct of human tasks.

Fire extended man's range and man's diet, thereby enlarging the world's carrying capacity for our species. Use of this form of energy set early hominids apart from other animals that relied entirely on their own metabolism. Fire's heat, used directly, helped make us human; but, in time, *Homo sapiens* attained a kind of *super*humanity by learning to convert the heat energy from fire into mechanical energy by means of various engines. Just before the Continental Congress gave the world a new nation to serve as the prime model of the exuberant way of life, James Watt devised for the world a practical engine for converting heat from fire into rotary motion, by means of steam pressure moving a piston in a cylinder and turning a crankshaft. The steam engine began to transform men into supermen. At first fueled sometimes with wood, later exclusively with coal, and still later largely with oil, these devices and their successors made available for the performance of an immense assortment of tasks the prodigious quantities of energy stored within the earth, in the transformed remains of organisms that lived millions of years ago. Man thus became uniquely differentiated from all other mammals by his ability to use solar energy that had reached the earth long before he existed—energy captured in prehistoric photosynthesis.

A geological savings bank had been accumulating these deposits of fossil energy for hundreds of millions of years. The steam engine and various improved successors to it—gasoline, diesel, jet, and rocket engines—gave man the key to this geological bank. What marvels poured forth from the turning of that key! The energy expended in two decades by a vast labor force of Egyptians stacking up some 2,300,000 blocks of stone (each weighing about two and a half tons) to form the Great Pyramid of Cheops was less than the energy released in a few *minutes* by three stages of a Saturn V rocket propelling men toward the moon.[14] Little wonder that the illusion of limitlessness was reinforced by space flights. In 1972, for example, the first Bangladesh head of state, Sheik Mujibur Rahman (destined for assassination within a few years by disillusioned constituents), pleaded for American aid for his newborn and frail nation, exclaiming, "If you Americans can put a man on the moon, you can do anything."[15] It seemed so, but it was far from true.

Fossil-fueled engines were to serve as the prime mover in launching the Industrial Revolution; they thereby brought about comprehensive reorganization of human societies. Industrialization compounded exuberance; James Watt was as much a revolutionary as Columbus had been. The apparent limitlessness of opportunities was underscored by the availability of energy in such quantities, and at such unprecedentedly low cost per man-hour equivalent, that human slavery lost its economic value. When men very much wanted tasks accomplished but could not or would not pay a free worker's wage, enslaving other human beings used to be the only solution. Clever machinery and cheap, abundant energy broke this pattern, serving as a great emancipator—the more so as invention continued to widen the range of tasks in which power-driven apparatus could substitute for human hands.

As we shall see, the "abundance" of this fossil energy was due to man's readiness to withdraw and spend it thousands of times faster than nature had deposited it in the earth's savings. And the energy from fossil fuels was cheap only because no workers had been paid (or slaves maintained) to grow the vegetation from which coal and oil had been formed. The cost of fossil energy, then, was determined essentially by the cost of extracting these fuels from storage. At the beginning of 1974, when oil prices had begun to rise sharply, a million kilocalories (equivalent to approximately one man-year of food-supported muscle power) cost only about $16 if obtained by burning 32 gallons of gasoline (at service station pump prices in the U.S.). That amount of energy would have cost some thirteen times as much if obtained from burning a liquid derived from contemporary agriculture (e.g., corn oil, 52 gallons at $4 per gallon, retail). If gasoline and other fossil fuels had been thirteen times as costly, we would never have fallen into the trap of reorganizing our social systems around their abundant use. Our overcommitment to dependence on fossil acreage was the result of the temporarily low cost of energy from antiquity. Because the low cost *was* temporary, it was an unrealistic basis for a way of life.[16]

Within two eventful centuries of the time when James Watt started us substituting fossil energy for muscle power, per capita energy use in the United States reached a level equivalent to eighty or so ghost slaves for each citizen. The ratio remained much lower than that in many other parts of the world. But, dividing the energy content of total annual world fuel consumption by the annual rate of food-energy consumption in an active adult human body, the world average still worked out to the equivalent of about ten ghost slaves per person. Otherwise stated, the average human being, whatever he might have done in a day with his own unaided muscle power, could now do about ten times as much by using his bodily energy mainly to direct the activities of mechanical servants using energy derived from fossil fuel combustion (i.e., from ghost acreage). More than nine-tenths of the energy used by *Homo sapiens* was now derived from sources other than each current year's crop of vegetation. Plants capture *contemporary* solar energy and produce combustible organic materials from inorganic substances. The fossil fuels, on the other hand, contain *prehistoric* solar energy, for they were geologically formed from organic materials produced by ancient photosynthesis. The exuberant way of life was now based, therefore, on a pattern of energy use that involved a high ratio between prehistoric energy and contemporary energy—a ratio that could hardly continue. Yet until the Arab oil embargo in 1973 this fact went almost unrecognized by news media, and there was little concern among the general public about the ratio's precariousness.

Recognition of the social significance of physical energy remained almost nil among politicians and social scientists until depleted resources began failing to meet persistently exuberant demand. But in a book called *Energy and Society*, whose enormous importance was insufficiently realized when it was published just a decade after the end of World War II, Fred Cottrell of Miami University in Ohio made clear the fact that "man can exist only where he is able to replace the energy he uses up in the process of living. He must regularly be in control of energy equal to or in excess of this minimum. A permanent deficit makes life impossible."[17] Full comprehension of the information and a thorough understanding of the reasoning in Cottrell's vital but neglected book would have shown the salience of Borgstrom's "ghost acreage" concept for the post-exuberant world. It was important to consider not just the food that keeps human bodies alive, but the energy of *all* kinds used by the mechanical extensions of man's bodily apparatus. Chapter 9 will explain *why* this was so important. For now, it is enough to be aware that, throughout the world, vast quantities of machinery driven by vast quantities of fossil energy had become indispensable for doing the things that had become part of human living during four centuries of exuberance.

Precarious Way of Life

Any nation that realized its self-sufficiency had fallen to less than 10 percent would almost certainly sense the precariousness of its existence. Borgstrom did not cite any nation whose visible acreage met as little as 10 percent of its needs. In energy terms, however, the condition of the post-exuberant *world* had become precarious in just that way. The human species, through technological progress, had made itself more than 90 percent dependent on phantom carrying capacity—a term we must now define. Phantom carrying capacity means either the illusory or the extremely precarious capacity of an environment to support a given life form or a given way of living. It can be quantitatively expressed as that portion of a population that cannot be permanently supported when temporarily available resources become unavailable.

Although the living generation did not realize that it was 90 percent redundant, the effects of dependence on phantom carrying capacity were beginning to be noticeable and disturbing. The reason for these effects remained unacknowledged, due to the continued grip of obsolete concepts on our thinking. Obsolete (i.e., ecologically naive) concepts impaled the minds of those in high office as well as the man in the street. As late as the end of 1973, both the president of the United States and the chairman of the Senate Interior Committee proclaimed as a goal of American policy the attainment of "energy self-sufficiency" by 1980. If the expression meant anything at all, it had to mean (in Borgstrom's terms) a goal of deriving all energy used by Americans from visible acreage, none from trade acreage. But the living generation could not become *really* self-sufficient just by ending its dependence on trade acreage; this would only accelerate the drawdown of energy deposits hidden beneath the domestic (visible) acreage. To achieve independence from OPEC opportunists by this method meant inflicting upon American posterity a legacy of aggravated resource depletion. In short, energy independence was illusory so long as massive quantities of energy were still to be obtained from fossil sources.

Neither the senator nor the president showed any understanding of the deep dependence of all modern civilization upon imports of energy from the prehistoric past. In 1970 American energy use amounted to the equivalent of approximately 58 barrels of oil per capita annually. Were it to become necessary to obtain all this energy from agricultural crops (i.e., from contemporary solar income, the only basis for permanent "self-sufficiency") rather than principally from the earth's savings deposits, the acreage required can be calculated as follows. Suppose alcohol derived from corn were to be the standard fuel. From each bushel of corn, about 2.3 U.S. gallons of alcohol could be produced.[18] In 1970 the entire United States corn crop came to about 4.15 billion bushels; this would have yielded about 9.67 billion gallons of alcohol—if we had been willing to forgo exporting any of the corn, or eating any of it, or feeding any of it to livestock. Since each gallon of alcohol has heat value equivalent to about 0.7 gallons of gasoline, this means the entire 1970 corn crop, converted to alcohol, could have supplied less than 7½ percent of that year's domestic demand for motor fuel! It would have supplied only 1.27 percent of total U.S. energy consumption. Even the *record* corn crop tabulated in 1976 (just over 6 billion bushels[19]) would have supplied less than 2 percent.

45

In other words, if we could miraculously increase corn yields *about fifty fold,* making 5,100 bushels grow on each tract of land now producing 100 bushels, we could eat our corn and have our fuel, too—free from dependence on depletable antiquity.

Make the merely optimistic assumption that we could perpetually hope to harvest 100 bushels of corn per acre, using energy *inputs* no larger than the 31 gallons of gasoline equivalent per acre that a 1944 estimate said were required to drive the machinery used in modern farming.[20] (Actually, the impressive rise in per-acre yields since 1952 up to a point where that 100 bushel figure is plausible has depended on further large energy subsidies in the form of heavy applications of synthetic fertilizers.) The energy cost of raising each 100 bushels of corn would amount, then, to the fuel derived from about 19 bushels. So the *net* fuel production would be based on no more than about 81 bushels per acre. Each acre would yield the net equivalent of almost 3 barrels of crude oil. To provide for the energy Americans were using in 1970 entirely by growing corn and converting it to alcohol, we would have needed just about 20 acres of good farmland per capita. But in 1970, the United States had just slightly more than *five* acres of farmland per capita—about half the nation's total area, and barely more than one-fourth of what it would take to meet American energy "requirements" from energy plantations converting contemporary solar energy into fuel. If all the farms in America had been devoted entirely to growing fuel-producing corn, and all could have yielded a net 3 barrels of crude-oil-equivalent per acre, the nation's human carrying capacity without ghost acreage would have been no more than 51 million persons. (It would have been appreciably less than that, actually, because presumably we would have wanted to use some of the farms to produce something to eat.)

As a drawdown-dependent nation, the United States was thus already relying upon fossil acreage four times as extensive as its total visible farm acreage. Our concern for the incidental fact that an appreciable and growing fraction of that fossil acreage was overseas and under the control of potentially hostile peoples was permitted to overshadow the more permanently significant fact that fossil acreage anywhere, and under anyone's control, was non-renewable. We were living on four parts of phantom carrying capacity for every one part of permanent (real) carrying capacity.

It should be clear, therefore, that the actual population of the United States had already overshot its carrying capacity measured by the energy-producing capability of visible American acreage. To achieve genuine self-sufficiency in energy by 1980, assuming a 1970 way of life but depending on visible acreage only, the population of this nation would have had to level off no later than 1880.

So the exuberant lifestyles of modern *Homo sapiens* were far more precarious than people realized. They could be practiced only as long as it was possible to continue extracting enough fossil fuels to maintain a high ratio of ghost acreage to visible acreage.[21] When two things put this high ratio in conspicuous jeopardy, some perceptive people began at last to sense the fact that continuation of the exuberant way of life was threatened. One signal was the build-up of pollution; accumulation of the combustion products from fossil fuels began to indicate that continued use in continuously increasing quantities posed real hazards to human health and survival—and to the health and survival of other organic species upon which humans depend. The other signal was increased difficulty of obtaining the fuels themselves; some of the most accessible deposits had been depleted, and some of the largest known remaining reserves were under the territories of nations not always eager to turn over such wealth to American or West European consumer nations—at least not without exacting a price that seemed exorbitant by standards forged in the Age of Exuberance.

That age was over, and its standards were already obsolete. Because of cultural lag, their obsolescence would be only belatedly recognized. Meanwhile, prices would inevitably rise. Politicians and pundits, working from the old paradigm, would continue invoking merely fiscal explanations for this inflation, neglecting its ecological basis. Among its effects would be some loss of the emancipating influence of cheap energy. Opinion leaders would generally continue seeking political explanations for the erosion of freedom, and would continue neglecting the ecological pressure causing it. By scorning as mere "Arab blackmail" the second signal mentioned above (the increased difficulty of obtaining fuels), such minds insisted on remaining blind to a reality far more significant than its surface political manifestations.[22]

Solutions That Aggravate Problems

Most of the world did not recognize the extent to which it was dependent on phantom carrying capacity in its use of fossil fuels. Non-recognition of dependence on invisible acreage, or the illusion of self-sufficiency, could lead to disaster, for actions based on illusions are inherently hazardous.

47

Consider, for example, the consequences of imagining that the resources of finite oceans were limitless. The more dependent a given nation became on "fish acreage," the more it was driven to improve the technology of fishing or to increase the fraction of its labor force engaged in fishing. It *needed* to maximize its proficiency in harvesting food from the sea. The more fish it could take, the better its people could be supported—as long as there remained fish stocks to draw down. When the oceans seemed vast and fish stocks seemed unlimited, there were no inhibitions against perpetually increasing the annual catch. By the time the danger of destroying the resource became evident, the people who needed the fish were already present, and the nation's dependence on resource-destroying rates of harvest was already established. Only after a lapse of time would calamity overtake it. If a fish-dependent nation's rate of harvesting fish exceeded the sustained yield rate, the effects of the damage to world fish stocks would be spread (for the time being) among other fishing nations. Although only a fraction of the immediate damage would be felt by a particular fisheries-dependent nation, that damage-committing nation would get *all* the benefit of its own excessive harvest.

Whenever the in-group directly and exclusively benefits from its own overuse of a shared resource but the costs of that overuse are "shared" by out-groups, then in-group motivation toward a policy of resource conservation (or sustained yield rates of harvesting) is undermined. In other words, competition for scarce resources is the enemy of self-restraint. This becomes especially so, as Garrett Hardin showed in a 1968 article in *Science* magazine, when scarcity becomes increasingly acute.[23]

Even so, in the case of "fish acreage," it was possible for substantial numbers of people both in the fish-dependent nations and in the more nearly food self-sufficient ones to *see* that a rate of harvesting in excess of the rate of replacement led to exhaustion of the resource in the long run, and could be advantageous only in the short run. Indeed, an appreciable fraction of the earth's human population was apparently beginning to grasp the sustained yield idea in regard to living resources such as fish or forests.

Understanding a principle and abiding by it are not the same thing. Overfishing continued in the 1970s, from necessity. The need for food *now* prevented men from always exercising the self-restraint they might know was necessary to ensure food for posterity. Posterity doesn't vote, and doesn't exert much influence in the marketplace. So the living go on stealing from their descendants.

Comprehension of the principle of sustained yield with respect to fish acreage has not prompted people to extend the idea very far into their thinking about other kinds of resources. Ghost acreage of the Carboniferous period was the resource base for "modern" living. In Asia, Japan was the nation most dependent upon such prehistoric photosynthesis. In Europe, Britain has been dependent on it longer than other nations. Americans were heavily dependent upon it, in spite of their huge expanse of visible acreage and their conspicuous agricultural surpluses. The more "modern" a nation had become, the more its way of life was based on importing energy from hundreds of millions of years ago. Yet so powerful was the old paradigm that it prevented most minds from even entertaining the thought that a replacement rate for fossil fuel deposits was as salient as the replacement rate for fish.

We have overused fossil acreage far more than we have overfished the seas. Overfishing means harvesting fish faster than they replace themselves by reproduction and growth. The inevitable result of continued overharvesting is eventual exhaustion of the resource. If it had been thought that men were harvesting fish as much as 1,000 (or even 100) times faster than the fish could replace themselves, there would have been much alarm around the world already. By 1970, the worldwide ratio between our use of fossil fuels and the natural rate of their replacement by geological processes was more like 10,000 to 1. Yet, at least until 1973, neither the ratio nor even the concept of replacement as such had entered the thoughts of most of the world's ravenous users of prehistoric energy, imbued as they still were with the notion of limitlessness.

For human minds shaped by the culture of exuberance, the closest approach to concern for a replacement rate in the realm of energy seemed to be the vague public awareness that continued geological exploration was constantly leading to discovery of additional deposits of fossil fuels. New discoveries would "replace" the known "reserves" mankind was continually depleting. Oil wells were known to run dry, but new wells were continually being "developed." The rate of discovery had no relation to the rate at which nature was laying down these deposits, but it *looked like* a rate of replacement. For a while it exceeded the rate of extraction (misleadingly called "production"), so the *illusion* of a sustained yield felt almost plausible, even for a resource that, in principle, had to be exhaustible. Deceptive language supported this illusion.

In the 1950s, however, in America's conterminous forty-eight states the discovery rate for petroleum deposits had peaked and turned downward. The downturn came in spite of improved technology for

49

exploration, expanded geological knowledge, and intensified effort—because only the less readily discoverable deposits remained to be found. When the "production" rate was less than the discovery rate, as it had been for several decades, the known "reserves" had been increasing from year to year (though the oil that *existed* in the world was not increasing). "Production" continued increasing even after the discovery rate began to decrease. The two curves crossed in 1961.[24] Thereafter, even the superstitious notion that a rate of discovery somehow sufficed in lieu of an actual replacement rate could no longer support the illusion of sustained oil yields.

Modern man misled himself in a number of ways. He made prodigal use of prehistoric ghost acreage to achieve illusory increases in "efficiency" in farming the visible acreage of the present.[25] Cottrell showed in his book, for example, that much more energy was invested to raise 50 bushels per acre in wet-rice farming by mechanized methods in Arkansas than by hoe culture methods in Japan. The illusion that the Arkansas procedures were more "efficient" arose from the fact that less than two man-days of human labor per acre were involved there, as against 90 man-days in Japan. To achieve that saving of 88 + man-days of human labor, however, the Arkansas farmer had to invest in tractor and truck fuel, electricity, and fertilizer, all involving energy that was equivalent to at least 800 additional man-days of effort. This energy extravagance would be even more flagrant if the accounting included the energy used in manufacturing, shipping, and maintaining the tractor, truck, electric pumping apparatus, etc.

Toward the end of 1973, when a no longer deniable shortage of petroleum was curtailing the use of automobiles in many countries, and was producing other unanticipated modifications of human activity, one American food distributor warned customers that food bills might be increased more than travel costs by the oil shortage. The distributor reported that the U.S. Department of Agriculture had said some 30 percent of the nation's fuel consumption was used in growing food and conveying it to the consumer's table. What neither that distributor nor his customers seemed to recognize was that the figure cited implied that several times as much energy went into producing, processing, and distributing food as the food itself contained! In terms of "newspeak," the perverted language from George Orwell's dystopian novel, *1984,* here was another inversion of meaning, similar to "war is peace" and "freedom is slavery." Fossil fuel use had enabled man to believe that "prodigality is efficiency."

Under these thoughtways men continued at the close of the 1970s to imagine that the solution to energy problems was to improve the technology for locating deposits and for extracting combustible substances from nature's underground storage, or to increase the financial incentives for doing these things. It was as if a family whose members were living far beyond their current income should urge the head of the household to solve their problem of overspending by increasing his proficiency in filling out withdrawal slips at the bank. It was as if they were to commend rather than reprimand him for withdrawing more each week than the week before. Newspeak: "Extraction is production."

Living on Ten Earths

A good estimate of the rate at which nature might be replacing the energy deposits man was withdrawing could have been easily calculated. One merely needed to know (1) the total weight of the earth's atmosphere, (2) the fraction of it that was oxygen, (3) how long it had taken for that much oxygen to be released from carbon dioxide (in which it had formerly been bound), and (4) the comparative weight of the one atom of carbon to the two atoms of oxygen in each former molecule of atmospheric CO_2. None of this information was secret or undiscovered; it wasn't even very obscure. Sea-level atmospheric pressure was commonly known, as was the approximate diameter (from which could be calculated the surface area) of the earth. So the weight of all the air on earth could be calculated to a reasonable approximation with ordinary high school mathematics. Roughly one-fifth of the air was now oxygen, and 99 percent of that free oxygen had been released, it has been estimated, in the last 600 million years.[26] The atomic weights of carbon and oxygen were readily available, and their ratio was simple to calculate. So it turned out that about 625,000 tons of carbon per year had been the average amount buried in deposits of coal, oil, natural gas, and other less combustible substances since the photosynthetic process began releasing into the atmosphere a net total of one million billion tons of oxygen. Much of that extraction of carbon from the atmosphere had occurred in the Carboniferous period, between 215 and 300 million years ago, so the *present* average annual addition to the world's fossil fuel deposits could scarcely be as much as half the long-term average.

By the 1970s, however, the world's human population, with all its technology, was burning these substances at a rate that re-oxidized and returned to the air more than four billion tons of carbon each year. In short, the rate of "harvesting" from this ghost acreage (4×10^9 tons per year) was more than 10,000 times what the rate of replacement might now be ($\frac{1}{2} \times 6.25 \times 10^5$ tons per year). Conservative as the estimate of a 10,000 to 1 ratio might be, it was not calculated in time to deter deep commitment of human societies to such overuse.

Even more simply, it would have been possible (had it not been for the pre-ecological paradigm) to see how much the output of agriculture and forestry and fishing would have had to increase if *Homo sapiens* were to try to derive more of his current energy expenditures from current energy income. Man was withdrawing annually from savings about ten times as much energy as he was obtaining from current income (from organic sources); therefore, to reduce his dependence on fossil acreage by only one-tenth, man would have to double his use of contemporary photosynthesis. That would obviously entail improvements falling somewhere in the almost surely unattainable range, between another doubling of yield per acre and another doubling of tilled acreage at existing yields.

To become *completely* free from dependence on prehistoric energy (without reducing population or per capita energy consumption), modern man would require an increase in contemporary carrying capacity equivalent to ten earths—each of whose surfaces was forested, tilled, fished, and harvested to the current extent of our planet. Without ten new earths, it followed that man's exuberant way of life would be cut back drastically sometime in the future, or else that there would someday be *many fewer* people. Neither alternative, and none of the reasons for them, were contemplated by those who glibly sought "energy independence."

James Watt had been conventionally regarded as something of a cultural hero for giving man access to a vast "new" source of energy. In the eighteenth century no one could recognize that, by inventing the steam engine, Watt was inventing a way for mankind to overshoot the sustainable carrying capacity of this one earth. Watt was a clever and decent man who lived in (and exemplified) the Age of Exuberance. His invention compounded the influence of Columbus's discovery, extending the carrying capacity surplus that briefly shaped our ideas, our lives, and our institutions. Watt reinforced man's belief in limitlessness.

Neither Watt nor Roosevelt, who also reinforced that belief, was ever taught to think in terms of carrying capacity or ghost acreage. So Roosevelt could not know, while inspiring and leading his countrymen toward recovery from economic depression, or when helping ensure Allied victory over Axis aggression, that he was prolonging unrealistic expectations of exuberance. No one in his "brain trust" could warn him of this, because even the keen minds of his advisors were tuned to the old cornucopian paradigm and were not trained to recognize the perils of dependence on phantom carrying capacity.

Once mankind was committed to heavy reliance on continued use of exhaustible resources such as the deposits of fossil energy, it was certain to be as painful for people to emancipate themselves from their own technological entrapment as it had been for earlier men to emancipate themselves from owning human slaves.

From the end of 1973, Americans began worrying about the ratio between their consumption of "foreign oil" and their consumption of "domestic oil." We let these worries overshadow completely the more profound issue that should have concerned everyone: the ratio between our dependence on energy from antiquity and our use of contemporary energy, i.e., the ratio between expenditures of withdrawn savings and expenditures of current income. The four billion human inhabitants of this one earth had learned to live as if they could count on harvesting each year the equivalent of ten earths' worth of combustible substance.

Notes

1. Recognition of this called for a change of national policy that long ago. See Whelpton 1939.
2. Compare discussion of the revolutionary potential when "cultural lags" pile up (Ogburn 1957) with the statement by Heilbroner (1974, p. 132; listed among references for Ch. 1) explaining why the outlook is for "convulsive change."
3. Frederick Jackson Turner, "The Significance of the Frontier in American History," *Proceedings of the State Historical Society of Wisconsin* 41 (1894):79–112.
4. Neither the change nor the misunderstanding of it would be exclusively American. This was important: the *world* had been affected by the free land in the Western hemisphere. It had also been affected by the technology that increased man's power to extract from the land in both hemispheres more wealth than earlier generations ever dreamed possible. So the whole world was now also affected by the *filling up* of formerly free land, and by the accumulated side-effects of modern technology. See Webb 1952 (listed among references for Ch. 2) and Cottrell 1955.
5. There were a few ways in which bigots might misread their racism into a book like Vogt's, and some writers (e.g., Allan Chase, *The Legacy of Malthus: The Social Costs of the New Scientific Racism* [New York: Alfred A. Knopf, 1976], pp. 378–380) were offended by these aspects of the book.

In a later book (*People! Challenge to Survival* [New York: William Sloane Associates, 1960]) Vogt acknowledged that he had been accused of racism for urging sharp reduction of birthrates especially among Latin Americans, Asians, and Africans, but he pointed out that his accusers "chose to forget my belief that the United States would [also] be better off with less people." They also seemed to have read through *Road to Survival* without grasping its central message, as expressed in statements like these:

> [p. 80] We must realize that not only does every area have a limited carrying capacity—but also that this carrying capacity is shrinking and the demand growing. Until this understanding becomes an intrinsic part of our thinking and wields a powerful influence on our formation of national and international policies we are scarcely likely to see in what direction our destiny lies.
>
> [p. 284] By excessive breeding and abuse of the land mankind has backed itself into an ecological trap. By a lopsided use of applied science it has been living on promissory notes. Now, all over the world, the notes are falling due.
>
> Payment cannot be postponed much longer. Fortunately, we still may choose between payment and utterly disastrous bankruptcy on a world scale. It will certainly be more intelligent to pull in our belts and accept a long period of austerity and rebuilding than to wait for a catastrophic crash of our civilization.

Critics who dismissed Vogt as an implicit racist were evading the necessity of facing that choice between revising our drawdown policies and undergoing global bankruptcy. Vogt had said (p. 284), "In hard fact, we have no other choice." And he was hardly being racist or xenophobic when he insisted (p. 285), "Drastic measures are inescapable. Above everything else, we must reorganize our thinking. If we are to escape the crash we must abandon all thought of living unto ourselves." By accusing Vogt of racism, however, preoccupied critics could even remain blind to such warnings as this:

> [p. 68] We are an importing nation; and every day we waste hundreds of millions of gallons [of gasoline]. . . . Our tensions find outlets in . . . traveling at high speeds that reduce the efficiency of our cars. We build into our automobiles more power and greater gas consumption than we need. We use the press and radio to push the sales of more cars. We drive them hundreds of millions of miles a year in pursuit of futility. With the exhaustion of our own oil wells in sight, we send our Navy into the Mediterranean, show our teeth to the U.S.S.R., insist on access to Asiatic oil—and continue to throw it away at home.

6. For example, American officials urge Saudi Arabia to keep oil output high to help stabilize the world economy in the face of shortages from other sources; the administration pushes through Congress a proposal for an Energy Mobilization Board with powers to "cut red tape" (i.e., by-pass

environmental protection legislation) when energy-related projects such as pipelines, oil refineries, synthetic fuel factories, etc., are at stake; the government "deregulates" natural gas and petroleum prices partly to "give incentives" to "producers."

7. Borgstrom 1965, p. 78.
8. Ibid.
9. See such sources as Small 1971; Colin Clark, "The Economics of Over-exploitation," in Hardin and Baden 1977, pp. 82–95; P. A. Larkin, "An Epitaph for the Concept of Maximum Sustained Yield," *Transactions of the American Fisheries Society* 106 (Jan. 1977):1–11. There is an important relation between the sustained yield concept and the concept of carrying capacity. Carrying capacity could be defined as the maximum population of an exploit*ing* species supportable by sustained yields of exploit*ed* resource species. See the definitions of these two terms given in the Glossary.
10. See Moorcraft 1973.
11. See Wynne-Edwards 1975.
12. See several of the papers in Schmidhauser and Totten 1978.
13. "Japanese Protest U.S. Fishing Limit," *Seattle Times,* Nov. 4, 1976, p. G4.
14. Richard S. Lewis, *Appointment on the Moon* (New York: Viking Press, 1969), pp. 504, 546, gives the total weight of the Apollo 11 Command Module plus Service Module plus Lunar Module plus Lunar Adapter as 50 tons. The velocity to which all this weight had been boosted when it left earth orbit en route to the moon was 24,000 miles per hour. It had thus had imparted to it 2.61 times 10^{12} joules of kinetic energy. For comparison: since three-fourths of the Great Pyramid's 450 foot height is above its center of mass, the 11.5 billion pounds of stone used to build it were raised an *average* 112.5 feet from the ground; this imparted to the 2.3 million stone blocks a total of 1.76 times 10^{12} joules of gravitational potential energy—roughly two-thirds of the energy imparted by rocket engines to the spacecraft bound for the moon.
15. *Newsweek,* Mar. 27, 1972, p. 39.
16. In 1979, American gasoline prices began to catch up with the higher prices most of the world's other peoples had already experienced for years. The rise continued to be mistaken for "gouging" or "blackmail," even though in ecological terms it was fundamentally an approach to greater realism, i.e., the beginning of a continuing correction of past underpricing.
17. Cottrell 1955, p. 4.
18. See Ayres and Scarlott 1952, pp. 233–239, and Cottrell 1955, pp. 141–142.
19. *Christian Science Monitor,* Nov. 12, 1976, pp. 1, 30.
20. Cottrell 1955, p. 142.
21. Recent research even indicates that biomass farming would, with present technology, yield *negative* net energy; energy inputs would exceed the

energy content of the usable fuels made from the harvests. This shows even more emphatically how dependent upon phantom carrying capacity modern nations have allowed themselves to become. See Weisz and Marshall 1979.

22. At a time when other nations were devising "carless day" schemes, or were at least having to curtail the hours or days of the week on which gasoline could be sold, the American *Daily News Digest* (put out by Research Publications, of Phoenix, Arizona) expressed its "conservative, free-market economics philosophy" by asking "Why is it that only the U.S. has a gasoline shortage?" The answer it suggested (in the third week of May, 1979) was that "only the U.S. has a Department of Energy." Startlingly similar views were expressed the following month by the 1976 winner of the Nobel Prize for Economics, Milton Friedman, who called for immediate abolition of the DOE and elimination of all price controls on petroleum products and natural gas, "confident that the market will promptly bring the energy crisis to an end." See his column in *Newsweek*, June 18, 1979.

23. Hardin's article has been widely cited and reprinted, and is included in Hardin and Baden 1977. See p. 28 therein.

24. See Hubbert 1969 (listed in references for Ch. 10), p. 178.

25. When used with insufficient care, "efficiency" can be a very misleading word. It always has a hidden reference: efficient with respect to what? In America and throughout the industrial world, *labor* efficiency has been purchased at the price of *energy* inefficiency. The latter type of efficiency has simply been neglected; as long as energy was unrealistically cheap and lavishly abundant, even the familiar concept of *capital* efficiency served very inadequately as a correlate or indicator of energy efficiency.

26. See the chapter by Lloyd V. Berkner and Lauriston C. Marshall in Brancazio and Cameron 1964. For a less technical version, see the article by the same authors in the tenth anniversary issue of *Saturday Review*, May 7, 1966.

Selected References

Ayres, Eugene, and Charles A. Scarlott
1952. *Energy Sources: The Wealth of the World*. New York: McGraw-Hill.
Billington, Ray Allen
1966. *America's Frontier Heritage*. New York: Holt, Rinehart and Winston.
Borgstrom, Georg
1965. *The Hungry Planet*. New York: Collier.
1969. *Too Many: A Study of Earth's Biological Limitations*. New York: Macmillan.

Brancazio, Peter J., and A. G. W. Cameron, eds.
1964. *The Origin and Evolution of Atmospheres and Oceans.* New York: John Wiley and Sons.

Bryson, Reid A., and Thomas J. Murray
1977. *Climates of Hunger: Mankind and the World's Changing Weather.* Madison: University of Wisconsin Press.

Cottrell, Fred
1955. *Energy and Society.* New York: McGraw-Hill.

Esposito, John C.
1970. *Vanishing Air.* New York: Grossman.

Hardin, Garrett, and John Baden, eds.
1977. *Managing the Commons.* San Francisco: W. H. Freeman.

Loehr, Rodney C., ed.
1952. *Forests for the Future: The Story of Sustained Yield as Told in the Diaries and Papers of David T. Mason, 1907–1950.* St. Paul: Minnesota Historical Society.

Loftas, Tony
1970. *The Last Resource: Man's Exploitation of the Oceans.* Chicago: Henry Regnery.

Moorcraft, Colin
1973. *Must the Seas Die?* Boston: Gambit.

Ogburn, William F.
1957. "Cultural Lag as Theory." *Sociology and Social Research* 41 (Jan.–Feb.): 167–174.

Ordway, Samuel H., Jr.
1953. *Resources and the American Dream.* New York: Ronald Press.

Park, Charles F., Jr.
1968. *Affluence in Jeopardy.* San Francisco: Freeman, Cooper.

Schmidhauser, John R., and George O. Totten III, eds.
1978. *The Whaling Issue in U.S.-Japan Relations.* Boulder: Westview Press.

Small, George L.
1971. *The Blue Whale.* New York: Columbia University Press.

Turner, Frederick Jackson
1920. *The Frontier in American History.* New York: Henry Holt.

Vogt, William
1948. *Road to Survival.* New York: William Sloane Associates.

Weisz, Paul B., and John F. Marshall
1979. "High-Grade Fuels from Biomass Farming: Potentials and Constraints." *Science* 206 (Oct. 5): 24–29.

Whelpton, P. K.
1939. "Population Policy for the United States." *Journal of Heredity* 30 (Sept.): 401–406.

Wynne-Edwards, V. C.
1965. "Self-Regulating Systems in Populations of Animals." *Science* 147 (Mar. 26): 1543–48.

4 | Watershed Year: Modes of Adaptation

Illusions and Delusions

Like Damocles, the courtier of ancient Syracuse who could not see what a mixed blessing it was to be king, we in the Age of Exuberance needed some reminder of the precariousness of high living. (Dionysius is said to have suspended his sword by a hair above Damocles' place at a banquet table to show him the sense of insecurity appropriate to a position of royal eminence.) We industrial hunters and gatherers, lacking any such reminder, were led by the culture of exuberance to suppose that mankind was largely exempt from nature's constraints.

This delusion of human exemptionalism was induced and encouraged by the myth of limitlessness made so plausible by the New World's carrying capacity surplus. Ancestors of today's Americans, in both the New World and the Old, acquired a deep conviction that men and women could live in abundance, working toward their own goals in their own way without interfering with each other's pursuit of happiness. That was the essence of the American dream, and that was what most fundamentally was changed by moving into a post-exuberant age. It was in America that the dream had seemed most obviously valid, so in some ways it was especially difficult for Americans to face the inescapable task of reassessing customary assumptions.

The old dream of an essentially non-competitive life (human brotherhood, we sometimes called it) could not be restored by militant activism, by revolutionary terrorism, or by dropping out of a corrupt social order. Nor could it be preserved by adamant resistance to political, economic, and social change. In response to post-exuberant distress, all of these strategies were tried in the 1960s and 1970s[1]; none could work, because the Age of Exuberance had been neither created nor destroyed by political action. Opportunities for non-competitive human relationships were inherent in the New World when it was new; they were irretrievably diminished when the New World ceased being new.

As humans became increasingly numerous, acquired increasing technological power, and faced diminishing natural resources, pursuit of self-interest no longer had the heroic ring of nation-building. Success in the post-exuberant age often came at the expense of others— not because we were more unscrupulous, but because we were more numerous and technologically more potent.

Modern technological advantages made us dependent on vanishing resources and on precarious economic relations. The world was becoming a place wherein actions that used to be quite harmless to others became harmful to all of us. Actions taken without any malevolent intention could now harm people by harming their habitat, or by using up resources not sufficiently abundant for all (or for posterity), or by giving unwarranted reinforcement to old illusions. Scientists breeding new high-yield strains of crop plants seemed to be fending off starvation, but the effect was to sustain population increase in an already overloaded world. Internationally, foreign economic aid turned out sometimes to be, in the final analysis, a means of enabling additional peoples to undertake living by drawdown, aggravating world dependence on phantom carrying capacity.

Progress also posed other problems. In medicine, new techniques of prolonging life thrust upon unprepared and unwilling human beings the burden of decision (rather than leaving to luck and microbes the task of "deciding") when human beings should die. New techniques of abortion began to give women "control over their own bodies" and freedom from "compulsory pregnancy"; they also began to give to the living more power to implement the antagonism between themselves and the unborn, antagonism which was being intensified by the age of population pressure. For the whole Western world, the tradition of bestowing upon future generations advantages exceeding those received from the past was becoming inverted. We had become competitors with, rather than benefactors of, our descendants.

A Specter Confronting the World

These were the circumstances in 1973 when the collision came. It was a head-on collision between the world's obstinate believers in limitlessness and the specter of too much technology wielded by too many inhabitants of a planet with obdurately finite carrying capacity.

The sense of having been hit hard by *something* began to evoke panic responses—a crashing decline in the stock market (unload your shares in energy-dependent enterprises while you can), spiraling inflation in consumer goods (grab all the profit you can before it's too late to do anything with it). Both these responses abated a year or so later, when it turned out that we had survived the collision for the time being. But the collision came about in such a way that adherents to the cornucopian paradigm could persist stubbornly in their non-recognition of *what* was hitting us. Smitten by the dependence of America and other industrial nations on prodigal use of exhaustible resources, opinion leaders grasped at the straw that appeared, and began to attribute the profusion of problems to an out-group. For yet another moment, the people could evade the realization that their own abundance was their basic affliction.

"Arab oil blackmail" was the phrase made to order for this evasion. Arabs obligingly presented themselves to the industrialized world as scapegoats. Conveniently, they were foreigners; blackmail was an unprincipled act of manipulation (with connotations of cruelty). As long as a sudden deluge of troubles could be attributed to villains in another land, the world could seem to remain in tune with traditional definitions of right and wrong. The industrial nations, their leaders insisted, had done a right thing by "developing" the oil resources of the otherwise impoverished Middle East. Arab leaders were doing a wrong thing in holding Europeans and Americans over a barrel and demanding that they recant their neutral or pro-Israeli attitudes.

When 1973 began, the belief that morality required firm resistance to oil blackmail was yet unborn, but resource shortages and other problems already beset mankind. A war-weary United States had only just extricated itself from its abortive military engagement in Vietnam; however, the end of American belligerency in Southeast Asia brought no end to the nation's woes. For a while the continuing troubles were imagined to be chiefly political—Watergate, abuses of power, and other scandals in Washington claimed public attention, nurtured cynicism in some, and reinforced in others the traditional expectation that a change of government would put things right. But lurking in the background were a host of economic pangs potentially more relevant as symptoms of the New World's superannuated state.

Over the celebrations of prisoner-of-war homecomings there hung a chill due to shortages of home heating oil. As spring advanced, shortages of various other commodities made themselves felt. Food prices increased throughout the winter and spring. Even government price freezes remained less than effective. One commodity in short supply was soybeans. "I can think of no shortage about which I could care less," I was told by one historian whose work had won him more than a dozen honorary degrees. To minimize internal effects of this shortage, the United States government banned soybean exports. The fact that Japan was heavily dependent upon American soybeans as a ghost acreage protein source was overlooked by the U.S. government (as well as by the learned historian). Japan protested. Amid some embarrassment, exports of soybeans to Japan were resumed, on a reduced scale.[2]

Then, as the 1973 wheat harvest got underway, growers were elated and bread buyers were outraged by the record heights to which wheat prices ascended. Both were still tempted to see the situation chiefly in political-economic terms, unmindful of the drawdown of world grain reserves. Sellers of wheat credited the administration for their sudden prosperity. Consumers of bread felt that administration-favored profiteers from the previous year's huge sale of American wheat surpluses to Russia were to blame for 1973's inflation in supermarket prices. Russia had needed American wheat to make up for weather-caused inadequacies in Russian wheat crops.[3] Americans imagined for a while that bad weather and crop failures could only happen in backward foreign countries, never in modern America. Thus we continued not to see that a world population dependent on optimum weather was a world population that had overshot the carrying capacity limits imposed by actual (fluctuating) weather.[4]

Meanwhile, summer tourists in various parts of America in 1973 were contending with service-station closures. Travelers occasionally were unable to buy gasoline by the usual tankful. Americans began to suggest government restriction of oil exports, not yet acknowledging the true dimensions of American dependence on *im*ports (already about 30 percent).

All these manifestations of the head-on collision *preceded* the sixth day of October, when Egypt and Syria again commenced their chronic hostilities against Israel. Chronology thus made it indisputable: the recurrence of conflict in the Middle East only aggravated but did not *cause* the resource shortages. Egyptian and Syrian forces gained a slight edge in the early days of fighting, due to the fact that, on the day when military action was launched, the Israelis were preoccupied with the Jewish day of atonement, Yom Kippur. By calling the 1973 outbreak "The Yom Kippur War," newsmen joined statesmen in letting themselves neglect a deeper reason for the Arab belligerents' choice of timing. The major powers that were ordinarily predisposed in favor of Israel were all located in the northern hemisphere and thus shared the same cycle of seasons. It was autumn for all of them when the Middle East hostilities broke out again. Over the past two decades these major powers had all made themselves more and more dependent upon oil,[5] much of it obtained from wells on Arab territory, not only for industrial uses but also for winter heating. In October, with winter just ahead, the leverage implicit in the strategy of oil export curtailment by Arab countries was approaching a seasonal maximum. It worked, making Israel's Arab opponents freer from constraint by the major powers than they might otherwise have been. Choice of Yom Kippur as the precise day of attack within the pre-winter season was perhaps only an added bit of tactical finesse.

As Egypt and Syria seemed at first to be forging ahead in a campaign to retake the Arab lands occupied by Israel since the six-day blitzkrieg of 1967, other Arab nations lent support or joined in. The 1973 Arab war effort also received encouragement and material support from the Soviet Union, giving the American government its customary rationale for taking sides. This enabled the United States to implement its traditional sympathy toward Israel by "resupplying" Israel's losses by a quick airlift of weapons and supplies—even though the dust had barely settled from the Asian war in which America had spent so much of its self-respect.

The U.S. government thus at first indulged its traditional pro-Israel reflexes—before starting in 1974 to make major concessions to the Arab nations. The administration in Washington was still *groping* its way into the new era. The Arabs had an ace up their sleeve, but Washington at first seemed unaware of it. Government minds apparently did not immediately envision the full economic impact of even marginal deprivation of oil. In no previous conflict had we been so dependent upon oil imports.[6] Nor did the men at the top see the futility of trying to counteract an oil embargo by indignantly stigmatizing it with a nasty label. "Blackmail" or not, the Arab pressure was going to work, because of the industrial world's need for ghost acreage.

Emergency sessions of the United Nations Security Council resulted in a cease-fire order. Pressure from their Soviet and American sponsors compelled the Arab nations and Israel to accede to this order, following only a few weeks of fighting so deadly that total Israeli losses were, on a per capita basis, comparable to those suffered by American forces in eight *years* of action in Vietnam. The cease fire was also preceded by a momentary threat of armed confrontation between the two superpowers themselves, as Soviet and American military forces went on an alert.[7] All this helped disguise the fact that, in the final analysis, by withholding oil for a few months the Arab nations succeeded in obtaining a truly drastic change in the previously pro-Israeli policies of the world's industrial nations, especially the United States.

When the sounds of battle subsided, the precarious coexistence in the Middle East continued. Maps were changed slightly to show Israel temporarily in control of somewhat more Arab territory, but the conservative King Faisal of Saudi Arabia was spearheading an Arab restriction of oil exports to Europe and a complete embargo on oil shipments to the United States. Faisal was later assassinated, but world dependence on Saudi Arabian oil continued.

In rapid consequence, the price of oil on the world market rose sharply. Even the Canadian government more than quadrupled its export tax on oil shipped to foreign customers such as the United States. Americans were suddenly obliged to acknowledge the fact that a nonnegligible fraction of the oil they had become accustomed to using was coming from outside their nation's borders. The ghost acreage concept was abruptly salient, though the phrase remained unfamiliar. People in all parts of the industrialized world became energy conscious. The previously unseen but chronic predicament of *Homo sapiens* came suddenly (and briefly) into public view as "the energy crisis." Where before there had been almost abysmal ignorance of the physical indispensability of energy as the basis of mankind's myriad activities, there was now astonished awareness. People were startled to discover the many ramifications of an energy shortage. Fear arose that a serious economic depression would soon befall the developed countries.

As 1973 ended, United States government policies and responses reflected an obstinate refusal to comprehend the full message about the perils of the drawdown method. As leaders of the world's most colossal industrial power, American officials chose to persist in the delusion that carrying capacity was unlimited. They sought to evade the limits imposed by the finiteness of man's global habitat, assuming that their constituents would suppose any felt limits were being imposed only by Arab governments.

For their part, Arab leaders seemed to sense that cleverness can be pushed too far and can become its own undoing. Too much pressure of resource deprivation could have aroused retaliatory reactions by the major oil-consuming nations—up to armed intervention in the Middle East "to protect our vital supplies" (and to uphold morality and punish "blackmailers"). Members of the administration in Washington let on that such a course of action was under consideration. However, in 1974 the Arabs, having won enormous concessions, turned the oil flow on again. In America many people did as their leaders had expected, by continuing to miss the point of what had happened. They persisted in discounting the significance of American dependence on phantom carrying capacity. The markedly higher prices they now had to pay for gasoline and fuel oil led many Americans to persuade themselves that the shortage had been contrived by oil-company tycoons for the purpose of price-gouging. The oil companies lent plausibility to this eager self-deception, for they took full advantage of the shortage to boost prices and reap sudden profits; they also launched huge advertising campaigns to extol their own devotion to serving the public interest through "developing" new (but more expensive) energy sources. Many common folk, like their government and like the ad writers, stubbornly denied the finiteness of man's habitat and the exhaustibility of its resources.

Repercussions

Obstinately insisting that the world was limitless did not remove its limits or keep them from having an impact. So, from this odd sequence of developments, fuels previously considered mundane and taken very much for granted began to seem precious. Assurance of their availability swiftly acquired priority over certain other values to which a great deal of lip service had been paid only a short time before. For example, the U.S. Congress enacted legislation in November 1973 to end the Environmental Protection Agency's ban on construction of the 800-mile oil pipeline across Alaska from the North Slope oil fields to the port of Valdez. A consortium of pipeline companies and oil firms had waited for years for the opportunity to invest several billion dollars in this project, only to be delayed by official and public concern for the hazards inherent in taking a fracturable pipeline across four earthquake zones and over permafrost that could be softened by the warmth of the oil flowing through the pipe, causing serious risks of sagging, breakage, and a massively polluting spillage of oil onto the tundra.

Similarly, the administration lifted its ban on oil drilling in California's Santa Barbara channel. Less than five years before, a Union Oil Company well some six miles offshore had blown out and spewed an estimated 3,000,000 gallons of oil into the sea. The well had flowed for eleven days before being capped, and something like $5,000,000 had to be spent afterward on clean-up operations. The incident had not only caused extensive damage to marine life and to coastal amenities; it had also aroused widespread opposition to offshore oil exploration and exploitation—overridden now by an unsatisfied "need" for fuel. Taking advantage of that need, the oil companies began pressing for authorization to drill offshore along the Atlantic coast, too.[8]

Various entrepreneur corporations had been waiting for oil prices to rise high enough to make it profitable to "develop" the vast oil shale deposits in Colorado, Wyoming, and Utah. Several of these firms now offered more money per acre to *lease* these public lands for this purpose than it cost to *buy* good farmland elsewhere in the country.[9] So-called development of this relatively unspoiled remainder of the American West would involve blasting mountains apart, scooping up the fractured rock, and cooking it to extract a barrel or so of heavy hydrocarbons per ton of rock. From this barrel of hydrocarbons could be made more or less the same products refineries had been making from easier-to-use liquid petroleum.

The fact that mankind now considered turning to this more difficult resource meant that the world's energy appetite had reached such a magnitude, and the surplus carrying capacity that had enabled mankind to enjoy an Age of Exuberance had been so drawn down, that people would now smash mountains in pursuit of their illusions. There was a pious pretense that natural landscapes would be restored afterward, but man's ability to imitate faithfully the natural effects of geological forces has never been great, and water shortages in the area, as well as skimpy financial arrangements, made the pretense appallingly transparent.

In response to the energy shortage, authorization was given for steam plants generating electricity to revert to burning dirtier fuels, or to operate at capacities that resulted in emissions exceeding anti-pollution regulations. A program of structural changes had been in progress when the oil shortfall became acute in 1973; additional steam plants were being converted from burning dirty fuels to burning comparatively less polluting fuels. This program was halted.[10]

To compensate for anticipated shortages of home heating oil as winter came on in 1973, the U.S. Department of Agriculture relaxed restrictions on cutting free firewood in its 155 national forests. Whereas previously only "bona fide settlers, miners, residents and prospectors" on national forest lands could indulge in this vestigial bit of frontiersmanship, now permits were granted without regard to an applicant's place of residence.[11] For some of the people who went into the national forests that winter with small chain saws and axes, taking home trailer-loads of firewood, the illusion of having retrieved the frontier era may have been briefly heartwarming. But no such lifting of controls could actually revive the Age of Exuberance, or restore the confidence in the future that was characteristic of a world with surplus carrying capacity.[12]

In Europe, more drastic measures were required, for European dependence on ghost acreage was more recognizable. European industry's postwar conversion from coal to petroleum had been almost as drastic as America's change; in addition, much more of the oil used in Europe came from external sources. Europeans rediscovered walking as a substitute for the automobile more readily than did Americans, and for a few months the governments of European nations were perhaps even less subtle than their American counterpart in making efforts to curry favor with Arab governments by abandoning Israel.

Through all these responses there ran persistent hope that the crescendo of shortages must be temporary. There was obstinate reluctance to face the fact that mankind was desperately dependent on ghost acreage, and that the world's human-dominated and technologically altered ecosystem was no longer operating on a sustained yield basis. Shortsighted preoccupation with the incidental fact that some of the phantom carrying capacity was based on imports from foreign places whose leaders might temporarily interrupt our access to it continued to obscure the larger fact that even more of it was being imported from the prehistoric past, and that nature would someday terminate the flow irredeemably. In an overpopulated and over-engineered world, the notion of limitlessness was quite obsolete.

Resistance to Realism

In mid-November 1973 a Philadelphia woman was quoted in an Associated Press article as saying there were "too many shortages for it to be real. Everything you go to buy they holler there's a shortage. Then they raise the prices and there seems to be an excess of it."[13] In New York City, another product of the culture of exuberance asserted his belief that "there is plenty of fuel." Of the national effort to cut fuel consumption, he complained: "I think it's a lot of politics. I just don't see the necessity."[14]

This kind of naive reaffirmation of the myth of limitlessness was not alien to ostensibly more sophisticated minds. For example, the editors of *The Economist* of London asserted in January, 1974, that "the recent wild changes in expectations about oil prices will lead to an energy glut . . . according to all that is known up to now about the elasticity of supply for energy."[15] They virtually echoed the views of the Philadelphia housewife when they set forth the basis of their expectation: "The present energy 'crisis' is about the fifteenth time since the war when the great majority of decision-influencing people have united to say that some particular product is going to be in most desperately short supply for the rest of this century. On each of the previous occasions the world has then sent that product into large surplus within 5–10 years." And they reflected the fact that technical knowledge of economic principles does not suffice to prevent mistaking the pace at which it is temporarily possible to draw down an exhaustible resource for the pace at which an industrial society can depend on using that resource. Like so many others, they let themselves be misled by conventional use of the word "production" to refer to a process of *extraction:* "In modern conditions of high elasticity of both production and substitution, plus surprisingly equal lead times for many investment projects, we now generally do create overproduction of whatever politicians and pundits 5–10 years earlier thought would be most urgently needed, because both consensus-seeking governments and profit-seeking private producers are then triggered by that commentary into starting the overproduction cycle at precisely the same time." Neither such faulty language nor implicit disregard for the world's finiteness can prevent oil wells from going dry. "Gluts" don't last.

From the following summer onward, the frequent violation of the fuel-conserving 55 mile per hour speed limit on American highways indicated that a good many people just did not and would not "see the necessity." But in Idaho, during the winter of 1973–74, the *Lewiston Morning Tribune* headlined a leading editorial more realistically: "The good life is running out of gas."[16] The editorial writer began by pointing out that population experts and environmentalists had been "wrong" in warning that serious shortages would occur in the 1980s and 1990s—wrong because the age of shortages arrived much sooner. After several paragraphs of comment on the seriousness of the oil problem, the editorial concluded:

> The sobering of America this year does not mean the end of the world. But it quite probably does mean the end of the very good life of unlimited consumption per capita on top of unlimited U.S. and world population growth.
> The Cadillac must give way to the Volkswagen, the bicycle, the bus and the foot. The toasty room will give way to the sweater. And there will be more sacrifices to come before, many winters hence, there will be some hope of fulfilling the lesser per capita demands of the smaller national and world population.

67

Less than six years later, as we continued to rely on phantom carrying capacity, the realism of this editorial was corroborated, and the temporariness of any plausibility for the "glut" hypothesis was revealed. Following the revolution in Iran, American motorists again competing for curtailed supplies of imported fuel actually resorted to occasional violence in gas station queues.

In Washington, D.C., in 1973–74, there was *unrealism* in the official stance of the administration. It was at this time that the president proclaimed the start of "Project Independence," a research and development effort he wanted Congress to authorize and fund so that the United States could attain "self-sufficiency in energy by 1980."[17] It was not widely recognized that such a project would only reduce American dependence on foreign-held ghost acreage by drastically increasing the already disastrous dependence on drawing down the fossil acreage beneath United States territory (and offshore). As the economist Paul A. Samuelson had shown, the follies of past energy economics included accelerated depletion of U.S. oil reserves by restrictive quotas on imports and by tax concessions to domestic oil extractors.[18] If the old paradigm had not had such a strong grip on so many minds, it should have been possible to see that Project Independence was mainly an inchoate plan for worsening American drawdown.

It was also hoped, of course, that "new sources of energy" would be "found" by Project Independence, including breakthroughs in nuclear power plant design, geothermal generating facilities, and technological devices for using solar energy. Such hopes amounted to a faith that endless substitutions of one resource for another could prolong limitlessness in the face of resource depletion. Like the family that lives beyond its income and expects to get away with it because next year the breadwinner's salary *may* be raised enough to cover the debts run up this year, the nation and mankind were increasingly staking their evasion of ecological bankruptcy on hoped-for but uncertain progress. (This evasive ploy will be examined further in Chapter 11.)

Tunnel Vision and Insight

This book is concerned with the predicament of the whole human species. But people differ, and it is important to recognize variations in human awareness of this predicament. Various combinations of insight and tunnel vision can be observed. Some people recognize that the New World is old, and that this necessitates major institutional change. Some people have faith that further technological break-throughs will obviate the need for major institutional change. Some people have faith that such measures as family planning, recycling centers, and anti-pollution laws will suffice to keep the New World new. Some people do not believe that the New World's former newness made any difference, or that its present oldness really matters. And some people stubbornly insist that the assumption of limitlessness was and still is valid.

These strikingly different opinions constitute different modes of adaptation to our post-exuberant age. Each represents a combination of responses to two new ecological understandings required by our new situation. First, people differ in their readiness to understand that the Age of Exuberance has ended, that this planet is already overpo-pulated, and that with our technology we have already made prodigal use of the world's savings deposits. Second, people differ in their readi-ness to understand that inexorable consequences follow from these changed circumstances (and that Arab leaders, corporate executives, or any other convenient scapegoats function only as *agents* of the un-avoidable change, not as its fundamental causes). All forms of human organization and behavior that are based on the assumption of limit-lessness will necessarily change *somehow* to forms compatible with the ecosystem's finite limits. The various combinations of responses to these two new ecological understandings are analyzed in Table 2.

The "Ostriches" in this analysis represent the truest adherents of the old cornucopian paradigm. The "Realists" are the truest adherents of the new ecological paradigm. The world looks very different to people who think in terms of such different perspectives. Communi-cation between them can be as difficult as between people who share no common language. Paradigm differences were thus bound to make painful the process of adjusting to post-exuberant realities.

Effects of Disillusionment

Transition from the old paradigm to the new one was complicated by events. Some deplorable things happened in the 1960s. With public attention riveted on events like the 1963 murder of President Kennedy and later preoccupied with eight years of unpopular war in Vietnam, it was easy to suppose that all the anxieties and antipathies so visible in that era were due to such events. It was not easy to see that the end of exuberance would have fostered dismal attitudes even if shots had

TABLE 2. Analysis of Several Modes of Adaptation to Ecologically Inexorable Change

	New Ecological Understandings			
ADAPTATIONS	CIRCUMSTANCE: The Age of Exuberance is over, population has already overshot carrying capacity, and prodigal *Homo sapiens* has drawn down the world's savings deposits.	CONSEQUENCE: All forms of human organization and behavior that are based on the assumption of limitlessness must change to forms that accord with finite limits.		NAMES
I. Some people recognize that the New World is old and that major change must follow.	= circumstance accepted	+ consequence accepted	=	Realism
II. Some people have faith that technological progress will stave off major institutional change.	= circumstance accepted	+ consequence disregarded	=	Cargoism
III. Some people have faith that family planning, recycling centers, and anti-pollution laws will keep the New World new.	= circumstance disregarded	+ consequence partially accepted	=	Cosmeticism
IV. Some people do not believe that the New World's newness once did, or that its oldness now does, have any significance.	= circumstance disregarded	+ consequence disregarded	=	Cynicism
V. Some people insist that the assumption of limitlessness was and still is valid.	= circumstance denied	+ consequence denied	=	Ostrichism

not been fired in Dallas and in the Tonkin Gulf. People nurtured in the culture of exuberance could hardly shift instantly, in the post-exuberant world, all the way from the Type V to the Type I attitude. Many were bound to become transitional types. Some made it to Type III (Cosmeticism) or Type II (Cargoism), but many others bogged down in Type IV (Cynicism). It was easy to misread that cynicism as merely a response to the war, or to the power structure's misdeeds.

The American people (and much of the world) had eagerly embraced the illusion, prevalent during the years of the Kennedy administration, that the so-called New Frontier, with its earnest drive to "get America moving again," was actually reopening the Age of Exuberance. That could not be done. But the will to believe had exceeded our ecological understanding. Disillusionment had to ensue.

Ecological reality made such disillusionment an eventual certainty even for perenially optimistic Americans. The world and its resources *are* finite, although we were trying still to live by a creed that denied this fact. Thus it was not the profound jolt from the first assassination of an American president in more than six decades that *caused* the chaos in the years that followed. That, and a divisive war, only *unleashed* the chaos (as some other jarring event might have done had these never occurred).

But what happened in 1963 did have *symbolic* significance. From Dallas onward, an abysmal sense of the mortality of all that is admirable led to a crescendo of ugly, mindless, and malicious behavior. The frustration-bred view that unattainable grapes must have been sour anyway was variously expressed. Grace and elegance, once admired, came to be widely detested as "establishment" values, to be mocked, or trampled under bare feet. The growth of worldwide "counterculture" movements, already begun, now accelerated. The neo-exuberant will to believe in a New Frontier was replaced in the later 1960's by an equally excessive will to believe that the old faith was not just innocently obsolescent but had always been flagrantly hypocritical. Naive but commendable Type V attitudes (such as the renewed idealism of the early 1960s that drew keen young Americans into the Peace Corps) gave way in a few short years to rampant iconoclasm, a version of the Type IV (cynical) attitude.

Typical Antinomianism

We need to recognize that all this iconoclastic defiance of once-respected norms was typical. Many times in previous centuries, in one part of Europe after another, the shattering of hope (by famine, conquest, or other disaster) had nurtured antinomian movements.[19] Participants in those movements would turn against the moral codes of their times, as if those codes, and not other circumstances, had been the source of their troubles. In the name of revolution or righteousness, they indulged in orgies or in masochism. Now this was happening again.

The significance of these responses would be missed if we did not discern that history was repeating itself. But it was so easy to be preoccupied with the specific grievances of the present and to overlook the recurrent patterns. It was so plausible to attribute the rage of the 1960s to frustrations arising from evil events of the 1960s. Patterned similarities between this and previous antinomian outbreaks were thus obscured, postponing for many of us our eventual recognition that the widespread sense of despair and loss of faith in the future were expectable repercussions of the post-exuberant condition of our world.

Reactions in the 1960s to the erosion of neo-exuberant hope ranged from ostentatious asceticism to "trashing" and wanton violence. These things were done in the name of various ideals: peace, love, equality, participatory democracy, law and order. Conventional wisdom attributed the loss of hope to factors as incidental as personality contrasts between the assassinated president and his successor,[20] or as accessory as the unresponsiveness of government, or the power of giant corporations. But America's troubles and the world's agonies were not that transient.

Antinomian behavior has been happening in many parts of the world, in nations that had not suffered a shocking presidential shooting, and in countries that were not party to an unwanted war.[21] Antinomian responses cannot re-create the carrying capacity surplus whose termination provoked them. Nor can any *attitude*. But any hope of *coping* with the difficulties inflicted by the carrying capacity deficit we now face must surely depend upon achievement of a *realistic* understanding of our situation.

Notes

1. For reviews and analyses of the tumult see O'Neill 1971; Roche 1972; Yinger 1977.
2. See articles in the *New York Times:* June 28, 1973, pp. 1, 77; June 29, pp. 49, 54; July 3, pp. 1, 41.

3. History repeated itself in 1979, when U.S. grain sales to the Soviet Union surpassed by 7 million metric tons even the 1972 figure of 18 million. See Ward Morehouse III, "New Grain to USSR, Higher Prices to You," *Christian Science Monitor,* Oct. 5, 1979.
4. For an indication of our dependence on a run of good weather, see Bryson and Murray 1977.
5. Davis 1978, pp. 48–109. See also Jensen 1970 (listed among references for Ch. 10) and Udall et al. 1974 (listed among references for Ch. 11).
6. See Schmidt 1974, pp. 213–221.
7. Ibid., p. 224.
8. Before the 1970s ended, Mexico had begun exploitation of huge oil deposits discovered in its coastal waters, the U.S. had sought to negotiate access to this new source of supply, and cartoonists had been blessed with an opportunity to satirize the way Mexican oil came to an oil-hungry U.S., after a Mexican offshore well went out of control and began repeating in enlarged form the Santa Barbara experience. Tarry blobs washed ashore on Texas beaches. The tourist industry on the Texas gulf coast suffered, not being imaginative enough to advertise for people to visit the area specifically for the purpose of having a look at "waves of the future."
9. Comparison based on local land prices in eastern Washington wheat country and shale leasing figures reported in *New York Times,* Jan. 9, 1974, p. 15.
10. See articles in the *New York Times:* Nov. 8, 1973, p. 51; Nov. 9, pp. 1, 26; Nov. 16, p. 21; Nov. 27, p. 40; Nov. 28, p. 1; Dec. 2, p. 71, and sec. IV, p. 5; Dec. 5, p. 37; Dec. 9, p. 70.
11. Reported by Associated Press, Nov. 19, 1973. See also articles in *Seattle Times,* Nov. 13, 1973, p. D3; *New York Times,* Dec. 22, 1973, p. 50.
12. In fact, by the end of the 1970s, woodburning stoves for home heating were reappearing on the American scene; see, for example, ads in various issues of *American Forests,* as well as mail order catalogs from Sears, Roebuck, and from other mail order houses. It began to appear that the U.S. would join the trend toward depletion of fuel wood supplies already causing serious problems in Third World nations; see Erik Eckholm, "The Other Energy Crisis," *American Forests* 81 (Nov., 1975): 12–13.
13. "Energy Crisis: Real to Many, Phony to Some," *Seattle Post-Intelligencer,* Nov. 15, 1973, p. D8.
14. Ibid.
15. "The Coming Glut of Energy," *The Economist* 250 (Jan. 5, 1974): 13–15.
16. *Lewiston Morning Tribune,* Nov. 19, 1973, p. 4. Cf. Ross Cunningham, "Upheaval in Our Way of Life Is Only Beginning," *Seattle Times,* Nov. 13, 1973, p. A12; James Reston, "Even Texas Is Running Short," *New York Times,* Nov. 14, 1973, p. 45.
17. Davis 1978, pp. 94–96.
18. Paul A. Samuelson, "Energy Economics," *Newsweek,* July 2, 1973, p. 65.
19. See Adler 1968; Cohn 1970 (listed among references for Ch. 11); Yinger 1977.

20. For example, Wicker 1968; Goldman 1968.
21. The reversion to Islamic fundamentalism in Iran after overthrow of the Shah in 1979 is a case in point.

Selected References

Adler, Nathan
  1968. "The Antinomian Personality: The Hippie Character Type." *Psychiatry* 31 (Nov.): 325–338.
Bryson, Reid A., and Thomas J. Murray
  1977. *Climates of Hunger.* Madison: University of Wisconsin Press.
Burch, William R., Jr.
  1971. *Daydreams and Nightmares: A Sociological Essay on the American Environment.* New York: Harper & Row.
Davis, David Howard
  1978. *Energy Politics.* 2nd ed. New York: St. Martin's Press.
Goldman, Eric F.
  1968. *The Tragedy of Lyndon Johnson.* New York: Alfred A. Knopf.
Marx, Leo
  1970. "American Institutions and Ecological Ideals." *Science* 170 (Nov. 27): 945–952.
O'Neill, William L.
  1971. *Coming Apart: An Informal History of America in the 1960's.* New York: Quadrangle Books.
Roche, George Charles III
  1972. *The Bewildered Society.* New Rochelle, N.Y.: Arlington House.
Schmidt, Dana Adams
  1974. *Armageddon in the Middle East.* New York: John Day.
Wicker, Tom
  1968. *JFK and LBJ: The Influence of Personality upon Politics.* New York: William Morrow.
Yinger, J. Milton
  1977. "Countercultures and Social Change." *American Sociological Review* 42 (Dec.): 833–853.

III | Siege and the Avoidance of Truth

Though progress has occurred in the past, its accelerating kinetics preclude it from being an everlasting feature of human history in the future.

> — Gunther S. Stent
> *The Coming of the Golden Age: A View of the End of Progress,* p. 94

The trends . . . do not point toward a sudden, cataclysmic global famine. . . . Marginal people on marginal lands will slowly sink into the slough of hopeless poverty. Some will continue to wrest from the earth what fruits they can, others will turn up in the dead-end urban slums of Africa, Asia and Latin America. Whether the deterioration of their prospects will be a quiet one is quite another question.

> — Erik P. Eckholm
> *Losing Ground: Environmental Stress and World Food Prospects,* p. 187

. . . domestic disturbances and slower [economic] growth rates will tend to force industrialized nations to cut back on their international commitments and be less generous in their foreign assistance.

> — Edward F. Renshaw
> *The End of Progress: Adjusting to a No-Growth Economy,* p. 235

. . . if the underdeveloped world is furnished with the mobile resources of capital and skill, its underdeveloped populations can be harnessed to modern production and modern technology and so enrich all nations to an extent unparalleled in man's records that would, within a measurable period, guarantee the abolition of poverty everywhere on earth.

> — David Horowitz
> *The Abolition of Poverty,* p. 165

Throughout their history Americans have insisted that the best was yet to be.

> — Henry Steele Commager
> *America in Perspective,* p. xi

5 | The End of Exuberance

Understanding New Circumstances

Until we learn to see the non-ideological reasons why the Age of Exuberance happened, and ended, we cannot comprehend our changed future. There were ecological prerequisites to abundance and liberty. Those prerequisite conditions are gone; in their absence, their social benefits may be impossible to perpetuate, at least on the scale we had come to expect.

The rest of the world can no longer be expected to attain the high standard of living to which so many in America became accustomed. Even where high levels of affluence have been reached, they are in jeopardy.[1] Insofar as democratic political structures were nurtured by the New World's carrying capacity surplus, the movement of mankind into an age of dependence on drawdown and an age of carrying capacity deficit threatens liberty.[2]

Pre-ecological interpretations of history, heretofore conventional, have tended to conceal the nature of our predicament. From the more revealing angle of vision developed in the first four chapters, therefore, let us now review the rise of a splendid American dream, take note of its ecological basis, and begin to assess the consequences that flow from termination of that ecological foundation.

The American Dream

Not so long ago, because the land and resources of the American continent seemed inexhaustible, men in both the Old World and the New fervently believed that America offered them unprecedented opportunities to succeed without having to put down rivals. The pioneer knew his life was a struggle, but he saw himself struggling mainly against the elements or against other species. Trees occupied land men wanted for other uses; unwanted plants and animals were ruthlessly

77

destroyed. (As we shall see in Chapter 7, this removal of species of indirect competitors to enlarge human carrying capacity was an important instance of mankind's perennial effort to turn back tides of ecological succession.) Some species were driven to the brink of extinction, or over it. The lordly bison, once so numerous, was almost wiped out. Predators such as coyotes and cougars were not tolerated, because they fed on the same species the pioneers fed on. The pioneer's struggle to master his new environment did sometimes include conflict with a race of *human* competitors whose invasion of the American land mass had antedated his own by twenty thousand years or so. However, the pioneer was no anthropologist; he was simply a seeker of life's blessings in what was to him a wilderness. Indians seemed a part of the New World's wildlife.

To put the matter ecologically, it seemed to the pioneer that competition in the New World was mainly between man and other species. In the Old World, with its carrying capacity for *Homo sapiens* (and his existing culture) already saturated, competition *within* the human species had been quite obviously a shaping condition of life. This difference between the two worlds, although partly illusory and somewhat exaggerated, mattered very much to settlers in America and to those in Europe who dreamed of coming to the land of opportunity. As the Chicago sociologist W. I. Thomas once noted, situations that men define as real have real consequences.[3] The American dream was a very real thing, even if in America's time of disillusionment it became fashionable to point out real discrepancies between the dream and our actual history.

From the beginning there was in America, as in all human societies, some hypocrisy, some ruthlessness, some subterfuge, some crime. On the frontier trapping, prospecting, homesteading, and nation-building were engaged in by rugged individualists—but there was also vigilantism. In this new nation blended from many nations there developed cultural pluralism—but there also arose attitudes of nativism and acts of lynching. There was slavery. Civil War came, then carpetbagging, then segregation and discrimination. It was useless for Americans to suppose, in their post-exuberant time of anguish, that these flaws in their heritage had never before been recognized. But just as truly it was misleading to forget (as many who developed Type IV attitudes of cynicism began to do in the 1960s) that earlier Americans really did expect wrongs to be righted. They expected change.

For a while it was genuinely believed that America was to be a "melting pot" which would fuse diverse peoples into a lustrous and durable alloy.[4] It was quite natural for Americans of an earlier time, so aware of their good fortune, to extend toward the Old World an open invitation for "The wretched refuse of your teeming shore" to come and participate in this new nation's good fortune. The more who shared in it, the greater it would become. The American dream held no premonition that someday the invitation would be withdrawn, perpetuated only as a poetic inscription on a bronze plaque on Liberty Island in New York harbor, commemorating a bygone era: "Send these, the homeless, tempest-tossed to me: I lift my lamp beside the golden door."

From the beginning of the nation to its bicentennial, some 47 million immigrants to America accepted that invitation. The fact that they kept on coming year after year and generation after generation confirmed the belief of those already living in America that it was a very special nation. Massive immigration was vivid testimony that the dream was shared by millions. Until recently that consensus was all the "evidence" most of us required to validate our dream. When we tried to express the dream in oratory, we continued to rely on words from Thomas Jefferson. Jefferson had not tried to be original; he only meant to express the aspirations widely shared by the people of the first new nation. It was to be a nation wherein all people could be recognized as equally entitled to life, liberty, and the pursuit of happiness. They could differ in *how* they lived, how they *used* their liberty, and how effectively they pursued happiness.

Basis of the Dream

Later, President Jefferson's purchase of the Louisiana Territory from France added to the United States an area larger than eight Great Britains. Jefferson explained his action to Congress by saying that this fertile and extensive country would afford "an ample provision for our posterity, and a wide spread for the blessings of freedom and equal laws." It was as if the author of the Declaration of Independence sensed that the human rights enumerated in 1776 could remain inalienable only if there could be perpetuated a fundamentally non-competitive relationship among men—by means of a "spread" wide enough to ensure low population pressure.[5]

With hindsight, it is possible to wish that an ecological vocabulary had been available soon enough for Americans of Jefferson's time and ensuing generations to understand unambiguously what their third president urgently sensed: a carrying capacity surplus facilitates development and maintenance of democratic institutions; a carrying capacity deficit weakens and undermines them. Had such knowledge been widespread, national policy might have given highest priority to perpetuating the condition of surplus and preventing transition to a condition of carrying capacity deficit.

When Americans not equipped with ecological concepts tried to describe and explain the contrast between their land of opportunity and the old countries from which they had come as immigrants, it became conventional to emphasize the political and ideological contrasts. We tended to forget that the freedoms America offered were not exclusively political. Even more, we forgot that the political differences between America and the older nations in Europe were caused by factors not mainly ideological. Europe was full of people; America was full of potential.

Ecologically speaking, the American dream expressed in human terms an exuberance that *characteristically* follows invasion of a new habitat by *any* species that happens to have the traits required for prompt and effective adaptation to it.[6] Human beings have exaggerated the apparent uniqueness of their own encounter with the felicitous circumstances of a New World. The "uniqueness of *Homo sapiens*" will be carefully reconsidered in a later chapter. For now, it is sufficient to note that few people have realized how frequently a similar experience has happened to other species whenever access was gained to a suitable but previously inaccessible habitat. As long as the members of an invading species remain far less numerous than the maximum population ultimately permitted by the carrying capacity of a new habitat, proliferation is easy, and competitive pressure upon the members of that species population will be low. Competition within the species may even be negligible when the small population is surfeited with unused resources waiting to be exploited.

It is time now to take note of the fact that the word "exuberance" has two meanings. It can refer to an emotional state, as well as to an ecological stage. The two meanings are connected. Some of the near synonyms for exuberance include such words as "luxuriance," "lavishness," "effusiveness," "superabundance," and "profusion." These words reflect the ecological meaning of exuberance—lavish exploitation of abundant opportunities. But other near equivalents can be "excitement," "ebullience," "effervescence," and "enthusiasm." Such words connote emotional exuberance, a mood that can approach euphoria.

What matters sociologically is that *the ecological kind of exuberance can produce among humans the emotional kind*. It did this for Americans, with profound social effects. Unused carrying capacity makes the future hopeful, which produces an optimistic spirit as people set about freely exploiting their abundant opportunities. America's history has thus exemplified the dependence of political liberty upon ecological foundations.

Both kinds of exuberance—ecological and emotional—had to be temporary, for they led inexorably to a change in the environmental conditions that had made them possible.

In 1896, before the United States had thrust itself into the role of an imperial power by taking overseas possessions away from Spain, a prophetic essay on this problem was published by the Yale sociologist William Graham Sumner. Sumner's parents had both been born in England. His father had migrated to the New World because of the scarcity of jobs in England and their abundance in America. Sumner had lived (as a graduate student) in Switzerland, Germany, and England, and had seen the difference between an old and crowded Europe and the spacious and vigorous America that was just beginning to become truly industrialized as he was growing up in it. Reliance on surplus carrying capacity acquired by takeover had only begun to be augmented (and was not yet overshadowed) by reliance on drawdown. Sumner entitled his essay "Earth Hunger."[7] Parts of it argued against the folly of political imperialism, suggesting that, for the already land-rich United States, an appetite for territorial aggrandizement motivated by national vanity would be seriously dysfunctional. But basically the essay was an incisive analysis of the dependence of democratic institutions upon favorable ecological circumstances. It was a warning few Americans chose to read, then or now.

The Unheeded Warning

The characteristics and well-being of human societies depend strongly upon the ratio between resources and people, as Sumner could see. This ratio limits the extent of civilization and comfort that can be attained. For various reasons, the social ramifications of the ratio can remain obscure. The ratio changes all the time. Nations change their boundaries. Even if boundaries don't change, human numbers do. Moreover, as we saw in Chapter 2, changing technology alters the utility of an acre of land. Though Sumner did not use ecological terms, he could see that a given quantity of land can have more carrying capacity as technology enables it to be worked more effectively. The advancement of knowledge increases man's ability to manage the forces of nature to his own ends. Developments of human *organiza-tion* also raise our power to extract abundance from a fixed quantity of land.[8]

The carrying capacity of a continent is thus neither fixed nor easy to calculate—but that does not make it infinite. For a given quantity of land and a specified state of knowledge and organization, Sumner pointed out that the population that can be supported must vary inversely with the standard of living. During an era of increasing productivity (when the population is well below the carrying capacity limit) this inverse relationship between population density and human well-being *seems* not to apply. In due course, it will.

All this was clear and seemed fundamental to Sumner, yet he could see that his countrymen tended to explain their standard of living and their liberty in quite different terms. America in 1896 was still in the exuberant phase. Sumner was ahead of his time in perceptively insisting that all forms of social disorganization and all human error have the effect of wasting carrying capacity—reducing the number of people who can be supported by a given acreage of land, or the level at which a given number can be supported. Most people had no incentive to think about such matters when there seemed to be carrying capacity to squander. All discord and quarreling, said Sumner, intensify the antithesis between population growth and per capita standard of living. The antithesis was invisible to most Americans in the Age of Exuberance because the experience of population pressure seemed so remote. It was invisible even to many Europeans, because America was there as a safety valve.

When population density is low, human equality is feasible and even probable. Each individual is economically valuable to others; it is, accordingly, hard for others to subordinate him. Class distinctions fade in such circumstances. Sumner tried to get us to see that democracy in Europe as well as in America had been fostered by the New World's low population density. This low density had relieved Old World pressure. Abundant land in another hemisphere influenced the European labor and land markets. Wages went up, food prices were held down, and land rents were kept lower than they would otherwise have been. The power of each European landed aristocrat was reduced by the availability of land elsewhere on the globe, not under his control. Still, Europeans tended to attribute their new freedom to new institutions and new doctrines, not seeing that the institutional and doctrinal changes were responses to the effective reduction of population pressure.

Long before Thomas Kuhn wrote about the perception-coercing

power of paradigms, Sumner pointed out how misleading doctrines legitimize abuses, provide alibis for human folly, slow down the machinery of society, waste effort, and reduce the effective carrying capacity of the habitat. The main reason we submit to such "sand in the gears," said Sumner, is because we have not acquired sufficient understanding of how things really work. Scientific inquiry, by disabusing us of doctrines that mislead us, can reduce wasted effort and make more effective our pursuit of happiness.

Sumner, who earlier had been a professional purveyor of doctrines—an ordained clergyman (in the Episcopal church)—came as a sociologist to regard the belief that prosperity and liberty were due primarily to ideologies rather than to low population density as not much different from the belief of the savage that a fetish worn on his person was what gave him success in hunting and fishing. Doctrines, Sumner now saw, could lower carrying capacity by fostering misuse of resources. (Nations may fight over doctrines, for example, and fighting or preparing to fight wastes resources. In the year when the American people celebrated the bicentennial of their independence, eighty years after publication of "Earth Hunger," the nations of the world—including the U.S.—were busily using an increasingly irreplaceable fraction of the planet's diminishing resources to support a global arsenal comprising 35,000 combat aircraft, 124,000 tanks, and over 12,000 combat ships. The total stockpile of nuclear weapons in existence added up to the explosive equivalent of about 12 tons of TNT for *each* of the earth's four billion human inhabitants.[9] If the *entire* gross national product of the United States for each of the first forty-six years of the twentieth century were expressed in 1976 dollars, the average per year would fall just short of the world's total military expenditures in 1976.[10])

Doctrines may be a frightful burden, Sumner said, for, with the prestige of antiquity and tradition, they deprive the living generation of an open-minded capacity to face facts.

Besieged and Bewildered

Some facts were almost too disturbing to face. Seventy-two years after Sumner's warning, the English novelist and physicist C. P. Snow journeyed through many lands around the world and reported that he found people in all of them feeling and acting like men under siege—anxious about the future, perhaps stoical, but not hopeful. As numerous and as technologically advanced as mankind on the average had already become after four centuries of exuberance, *we had besieged ourselves*. Each of us was now doing damage, just by living, to the life support systems of our finite planet.

It was almost literally *incredible* to people living two centuries after Jefferson that the world and man's relation to it had changed so fundamentally. The New World was not new anymore. Even our most normal and non-reprehensible ways of using resources to support human life and pursue human happiness were now putting our habitat and our species in peril. The Age of Exuberance was past. The world was now suffering the consequences of overshoot. The inverse relation between numbers and well-being was making itself harder to ignore. Ecological stresses gave rise to scarcity, inflation, and unemployment, and these economic conditions led to hunger, forced migration, and political unrest.

But many world leaders persisted in their commitment to a cornucopian paradigm. For example, the pope, to no one's surprise, during the World Food Conference in Rome at the end of 1974 insisted, "It is inadmissible that those who have control of the wealth and resources of mankind should try to solve the problem of hunger by forbidding the poor to be born."[11] From this statement it was evident that "carrying capacity," "overshoot," and "drawdown" were alien to the papal vocabulary.

Equally anachronistic and unsurprising was the vice-minister of agriculture from the People's Republic of China, who alleged at the conference that the industrialized countries had provoked the world food crisis.[12] (His vocabulary was ideological and did not equip him to see—and to say—that if the industrialized countries had done this, it was by committing themselves and the rest of mankind to dependence on ghost acreage; his concern was not the drawing down of temporary carrying capacity.) When he praised the oil-exporting nations for having achieved a "victory" over the industrial nations, he showed that the level of political power struggles was as deep as his thinking could go.

The Ostrich response of non-ecological minds was evidently not confined to frustrated consumers who thought there were too many shortages for the scarcity to be real. Dean Rusk, secretary of state under Presidents Kennedy and Johnson, also viewed the world in that way. Childhood poverty had compelled this future Rhodes scholar, college dean, army colonel, foundation head, and secretary of state to attend school without shoes. From the perspective of his own upward mobility, he saw the county where he grew up as an underdeveloped area that became modernized, and he generalized its attainments as the potential destiny of all underdeveloped areas. According to David Halberstam, in his own lifetime Rusk had seen that boyhood environment revolutionized "with education, with technology, with county agents, and with electricity." Rusk refused to believe assertions that underdeveloped countries would need two or three more centuries to rise to the American level of affluence "because I know it isn't true. Because I've seen it with my own eyes."[13]

Missionaries of Demoralization

The destructive potential of that kind of unrealistically expectant attitude had already been discerned. The Columbia University sociologist Robert K. Merton, himself a beneficiary of similar upward mobility, had pointed out in his 1949 article on "Social Structure and Anomie" that there is pressure toward deviant behavior on a large scale when a system of cultural values extols above all else certain success-goals *for everybody*, even though there are structural barriers restricting or precluding access to those goals by approved means for a considerable part of the population.

Under post-exuberant conditions, Secretary Rusk's Horatio Alger view of things typified a worldwide force that has fostered anomie. There have been efforts on all continents to attain, *by fair means or foul*, the blessings of limitlessness in the face of limits. As Merton had seen, exaltation of a goal that is unattainable by socially acceptable action generates demoralization—a "de-institutionalization of the means."[14] Accordingly, in a world where success for many had become unattainable because there was no more unused carrying capacity, a missionary effort to induce peoples of underdeveloped countries to aspire toward economic development should have been expected to lower both morale and morality.

Faced with this unintended and deplorable result of their missionary efforts, many of the benevolent people of the developed countries (who had held out the unreachable carrot to frustrate the underdeveloped nations) remained Ostriches (the Type V adaptation) or became Cynics (Type IV).

Achieving Chaos

The illusion of limitless opportunity for economic development persisted partly because it was embedded in such documents as the

United Nations Charter. According to the Charter preamble, the peoples of the United Nations were determined in 1945 not only to "save succeeding generations from the scourge of war" but also to "promote social progress and better standards of life in larger freedom" and "employ international machinery for the promotion of economic and social advancement of *all* peoples" (italics added). If the earth's limits made it fatuous to universalize these hopes, then (by reasoning along the lines set out by Merton) it followed that sentiments expressed in the Charter preamble were a prelude to global anomie.

But men of good will pressed on toward unsought chaos. In December 1948 the United Nations General Assembly adopted a *Universal* Declaration of Human Rights. Its preamble again proclaimed, among other things, determination "to promote social progress and better standards of life," and its Article 25, Section 1, said (italics added): "*Everyone* has the right to a standard of living adequate for the health and well-being of himself and of his family, including food, clothing, housing and medical care and necessary social services, and the right to security in the event of unemployment, sickness, disability, widowhood, old age or other lack of livelihood in circumstances beyond his control." Overshoot and resource depletion were certainly circumstances beyond the control of inhabitants of the post-exuberant world; the Universal Declaration virtually proclaimed their irrelevance as determinants of a human being's life chances.

Unseeing, and therefore undaunted, the General Assembly in 1960 adopted a Declaration on Granting Independence to Colonial Countries and Peoples. Among its ecologically questionable pieties was the proclamation that *all* peoples could "freely determine their political status and freely pursue their *economic,* social and cultural development." Inadequate "political, economic, social or educational preparedness," it said, must never serve as "a *pretext* for delaying independence" (italics added). By implication, economic inadequacy was expected always to be a trumped-up excuse and never a genuine reason for delaying independence. Again the very possibility that real limits might impede attainment of yearned-for goals in a post-exuberant world was not acknowledged, and apparently not seen.

In 1962 part of a former Belgian UN trusteeship in east-central Africa became a new nation, Burundi. A nearby former British protectorate gained independence as Uganda. But high hopes and benevolent intentions at the United Nations could not insulate mankind from travesty. The constitution of Burundi alluded to inspiration provided by the Universal Declaration of Human Rights, and in the Ugandan constitution there was a bill of rights, also traceable to its influence.

Subsequent inhumanities of man toward man in those fledgling nations seemed to corroborate the expectation derived from reasoning in the manner of Merton: to exalt goals that unalterable circumstances make unattainable is to play with dynamite.

Foreign economic aid had by this time become a more or less fully institutionalized expression of the conviction that, with enough help, underdeveloped countries could rise to industrial stature. Net official disbursements by the member nations of the Development Assistance Committee of the Organization for Economic Cooperation and Development more than doubled in a dozen years. Paul G. Hoffman, United Nations Development Program administrator, believed at the beginning of the 1970s that a solid foundation had been laid during the 1960s "for building new and better economies and societies throughout the low-income world." Means were at hand, he felt, "to complete this vital process of nation building. What remains to be tested, however, is the *will* . . . of all countries, rich and poor, to cooperate fully and to work unremittingly for meaningful development progress."[15]

An old cliché had asserted, "Where there's a will, there's a way." It was plausible when resources were undepleted and constraining pressure from non-volitional limits was negligible. But it was a manifestation of the obsolete cornucopian paradigm and could hardly suffice as a valid perspective for statesmen in the post-exuberant world.

Such pre-ecological thoughtways prevented most people from seeing that the United States and other industrial nations could not really export, in full measure, their own success. But they had already succeeded in exporting to the non-industrial peoples the *aspiration* to become economically developed beyond the world's capacity. Willard L. Thorp, who was for five years chairman of OECD's Development Assistance Committee, wrote in 1971 that economic growth had become a goal for which governments were expected by their constituents to strive. This was especially so, he said, for governments in the less developed countries. Leaders in those nations typically promised more economic growth (as an expected result of newly won political independence) than any regime could deliver. Many of the new governments were soon ousted because of "the inability of any particular set of incumbents to meet popular demands for specific improvements or a generally higher level of living."[16]

Under such circumstances no one should have been surprised at an outburst of genocide in one of the newborn African nations (Burundi), or, almost next door (in Uganda), the rise of a despot to power and his expulsion of persons of Asian descent from that country to enable black Ugandans to take over the property and the elite jobs of the expelled. These perversions of exuberant expectancy were the sort of things Merton's analysis of the anomie pattern made broadly predictable.

Whether Recognized or Not

But real limits not seen are not limits repealed. By whichever paradigm we comprehend things, the fact remains that it took ships, as well as desires, to bring European emigrants to the New World. It took abundant natural resources, not just human aspiration, to make the New World modern. (It would have taken unprecedented mutual restraint, as well as industrial progress, to make alabaster cities gleam, undimmed by human tears.)

The world's limits would limit, whether recognized or not. Ostriches in the State Department or in United Nations agencies, bound by paradigms from the Age of Exuberance, were sincere in their denial of nature's limits; nevertheless, they were doing a grave disservice to the peoples they meant to assist. The new ecological paradigm was prerequisite to seeing that *universal development was an unattainable goal*. It was tragic that such a goal came to be universally sought. The effects of this pursuit of the infeasible were grievous, and they were grievously misunderstood. Most of the poor nations of the world would never become rich. And the rich nations were just beginning to discover how improbable it was that they would remain rich in a poor world.

A mood like that of people under siege therefore overwhelmed the time-honored capacity of Americans for witty self-assurance. There was a grave loss of confidence in the future. This was not due to mere ideological apostasy; rather, the traditionally ebullient belief in the inevitability of progress was a casualty of our belated awakening to post-exuberant realities.

No inspirational rhetoric will suffice to revive a faith in the future which past generations drew less from fine prose than from the New World's once vast carrying capacity surplus. Persistence of an obsolete paradigm caused people to discount the fact that for no species, not even for man, can an Age of Exuberance be more than temporary in a finite world. But let us be relativistic enough to hold no grievance against our forebears, who in their day could not anticipate the pressures to which we, their descendants, were being committed by their great expectations.

Living in the twilight of the American dream and of the exuberant way of life is inevitably painful. The crowded world easily becomes a world in which antagonists engage in round after round of mutual "retaliation." Acts of terrorism beget acts of terrorism. But bitterness serves no purpose, even in these discouraging straits. Nostalgia is a more useful response. We should never disdain to cultivate memories that nurture courage and compassion. The right kind of nostalgia can be a legitimate and rational response to irreversible change for a nation or a world, just as it is for an aging individual.

Americans have perhaps been hit harder, in some ways, by the shattering of the dream, because Americans had experienced the greatest exuberance. By cultivating an ennobling form of nostalgia, perhaps we can prevent the blood spilled in Dallas and in Memphis, or the oil spilled off the Santa Barbara shore, from obliterating our vision of a nation undimmed by human tears, with brotherhood from sea to shining sea as a crowning expectation.

We have begun to sense that we are never going to achieve that utopian condition, but when reality falls short of our ideals we need more than ever to recognize that ideals have value in themselves as expressions of human nature.

We must not forget.

Settlers in a New World *did* create a new and inspired form of government in a land of opportunity. That land *did* provide a haven for millions of dispossessed, and a refuge from religious and political tyranny. Americans tore themselves asunder to recant the odious doctrine that man could be man's chattel, and painfully they sutured the national wound. Americans *did* win the West, and it was a West that far outshone its television parody. A great nation was built in a wilderness—and then its people learned to cherish samples of the wilderness before it was all changed. The nation tried honestly and passionately to preserve three dozen national parks and scores of national monuments as places of natural splendor and reminders of a pioneering heritage. It tried honestly and generously to share the fuits of its frontier experience with people in other societies overseas, where the national park idea was often genuinely appreciated when some of America's other cultural exports were regretted.[17]

Until recently, nothing required the citizens of an exuberant nation to realize that a pioneer community cannot endure forever without turning into something very different. It is not a "climax community." The meanings of such terms, and of the inexorable process of change called "succession," will be made clear in the chapters that follow.

Notes

1. Ehrlich and Ehrlich 1974.
2. That this would happen was seen as long ago as 1896 by William Graham Sumner, whose essay on the subject is discussed later in the present chapter.
3. The principle was quoted and elaborated upon by Merton (1968, pp. 475ff). Merton's interpretation of the Thomas principle may too easily be misconstrued as an assertion that the *actual* consequences of defining a situation in a certain way are inevitably just what the definition of the situation leads people to *expect*. This is not so (and it is not really what Merton was saying). All that the Thomas principle really tells us is that *it matters* how people define the situations they encounter. It does not declare that objective physical reality has no effect on human experience; nor does it mean that the consequences of a real situation depend altogether on the vagaries of expectation and perception. As Merton pointed out (p. 485), "Eventually, life in a world of myth must collide with fact in the world of reality."
4. For reference to this belief and discussion of its more recent erosion, see Nathan Glazer and Daniel P. Moynihan, *Beyond the Melting Pot* (Cambridge: MIT Press, 1963); and J. Milton Yinger, "Ethnicity in Complex Societies: Structural, Cultural, and Characterological Factors," Ch. 11 in Lewis A. Coser and Otto N. Larsen, eds., *The Uses of Controversy in Sociology* (New York: Free Press, 1976), pp. 197–216.
5. See Peterson 1970; Benson 1971.
6. See, for example, Quick 1974, pp. 31ff.; A. J. Nicholson, "An Outline of the Dynamics of Animal Populations," *Australian Journal of Zoology* 2 (May, 1954): 9–65.
7. In the same volume of Sumner's essays that includes "Earth Hunger" there is another that has gained increasing pertinence for understanding our post-exuberant predicament. Originally published in 1887, "The Banquet of Life" argued that if there were any such thing as was connoted by that then-modish expression, we needed to understand that it was "not set for an unlimited number."
8. Technology and organization are two of the three components of what Robert E. Park termed, in a 1936 article, "the social complex." The three components, in mutual interaction with each other and with an environment, constitute what Otis Dudley Duncan later called "the ecological complex." See pp. 251–252 in Riley E. Dunlap and William R. Catton, Jr., "Environmental Sociology," *Annual Review of Sociology* 5 (1979): 243–273.
9. John Dillin, "World-at-Peace Arms Cost—$350 Billion and Growing," *Christian Science Monitor,* May 9, 1977, p. 1.
10. Calculated from data given in Bureau of the Census, *Historical Statistics of the United States, Colonial Times to 1970*, Bicentennial Edition (Washington: U.S. Department of Commerce, 1975), I, 224.

11. Quoted in the *New York Times,* Nov. 10, 1974, p. 1.
12. Mentioned in a radio newscast. Cf. Gladwin Hill, "Marx vs. Malthus: Ideas Stir Rancor at Population Meeting," *New York Times,* Aug. 26, 1974, p. 10, and the editorial, "Demagogy in Bucharest," ibid., p. 28.
13. Quoted in Halberstam 1972, p. 314.
14. Merton 1968, p. 190.
15. Quoted from Hoffman's "Introduction" to Legum 1970, p. xxvii.
16. Thorp 1971, p. 25.
17. See the remark by a New Zealand visitor to some of the national parks in the U.S. during the Vietnam war, quoted in P. H. C. Lucas, *Conserving New Zealand's Heritage* (Wellington: A. R. Shearer, Government Printer, 1970), p. 84.

Selected References

Becker, Carl
   1922. *The Declaration of Independence: A Study in the History of Political Ideas.* New York: Harcourt, Brace.
Benson, G. Randolph
   1971. *Thomas Jefferson as Social Scientist.* Rutherford, N.J.: Fairleigh Dickinson University Press.
Ehrlich, Paul R., and Anne H. Ehrlich
   1974. *The End of Affluence.* New York: Random House.
Halberstam, David
   1972. *The Best and the Brightest.* New York: Random House.
Halderman, John W.
   1966. *The United Nations and the Rule of Law.* Dobbs Ferry, N.Y.: Oceana Publications.
Legum, Colin, ed.
   1970. *The First UN Development Decade and Its Lessons for the 1970's.* New York: Frederick A. Praeger.
Merton, Robert K.
   1968. *Social Theory and Social Structure* Enlarged ed. New York: Free Press.
Morison, Samuel Eliot
   1965. *The Oxford History of the American People.* New York: Oxford University Press.
Peterson, Merrill
   1970. *Thomas Jefferson and the New Nation.* New York: Oxford University Press.
Quick, Horace F.
   1974. *Population Ecology.* Indianapolis: Pegasus/ Bobbs-Merrill.

Sigmund, Paul E.
    1967. *The Ideologies of the Developing Nations*. Rev. ed. New York: Frederick A. Praeger.
Snow, C. P.
    1969. *The State of Siege*. New York: Charles Scribner's Sons.
Sumner, William Graham
    1887. "The Banquet of Life." Pp. 217–221 in A. G. Keller, ed., *Earth-Hunger and Other Essays*. New Haven: Yale University Press, 1913.
    1896. "Earth Hunger or the Philosophy of Land Grabbing." Ibid., pp. 31–64.
Thorp, Willard L.
    1971. *The Reality of Foreign Aid*. New York: Frederick A. Praeger.

IV | Toward Ecological Understanding

. . . human life represents a continuous process of metabolism in receiving energy, transforming it, and expending it in the form of excretions. . . . [This] process . . . ought to receive attention not only from the biologist but also from the sociologist, i.e., from the investigator of human behavior and of social processes.

> — Pitirim A. Sorokin
> *Hunger as a Factor in Human Affairs,*
> pp. 3–4

Whereas other animals depend largely on genetic changes for adaptation to environment, man's chief form of adjustment has been through agencies external to himself but largely of his own fashioning. Instead of developing claws, wings, hard shell coverings, horns, etc., man has constructed tools, clothing, weapons, and various other devices from the materials of his environment.

> — Amos H. Hawley
> *Human Ecology,* pp. 24–25

A pond in an evergreen forest may receive very little sunlight, deriving its energy from forest detritus that blows or falls into it. . . . For some purposes it is useful to conceptualize a city as a detritus ecosystem, dependent on external sources of energy and other resources.

> — Arthur S. Boughey
> *Man and the Environment* (2nd ed.), p. 9

. . . industrial technology is by nature exploitative and destructive of the materials that are necessary to maintain it . . . and at the very minimum must find and develop substitutes for those resources that become exhausted.

> — Richard T. LaPiere
> *Social Change,* p. 530

Eventually they cannot survive in the conditions they themselves have made.

> — part of a picture caption in an interpretive display about ecological succession, in the visitor center near Fishing Bridge, Yellowstone National Park

6 | The Processes That Matter

Escape from Arrogance

When human history did not work out according to the high hopes engendered by the Age of Exuberance, we began in the post-exuberant world to suffer from the illusion that we could retrieve the good old days by hating someone for stealing them from us. Examples abound. People to be hated have ranged from communists to capitalists, from whites to blacks, from true believers (wrong brand) to revisionists. Parties, nationalities, races, classes, and faiths have all served as targets for frustration-bred hatred. Weimar Germany became the genocidal Third Reich of Adolf Hitler because there were desperate or humiliated crowds willing to embrace the doctrine that "non-Aryans" were the source of their nation's woes.

Because the rest of us are susceptible to the same kind of enraged response to desperate circumstances,[1] we urgently need to see that what has befallen mankind was no mere manifestation of some sinister conspiracy. Nor was it due to some chronic flaw in the national character of any of us. It was the natural outcome of a natural process. We might endure it less self-destructively if we can learn to see it that way. Man has been too arrogant in exaggerating the difference between himself and other creatures, between human history and natural history.

According to the ecological paradigm this book is meant to advocate, what has happened to the American dream, and what is happening to human life all over the world, can be understood as ecological *succession* (and related processes). Members of the human species failed to heed clear omens of our own predicament because we did not know they were precedents. We had not yet learned what processes matter most. Some of the processes most deeply affecting human life were those that seemed trivial by conventional standards of thought. We disregarded innumerable instances in which populations of organisms so changed their own environments that they undermined their own lives. Most of us didn't know that this was what these examples exemplified. And we never supposed the pattern could include us.

In winemaking vats, which provide a finite but lavishly stocked habitat for micro-organisms that ferment juice from crushed fruit, the organisms grow exuberantly at first, deplete the initially abundant nutrients upon which they thrive, and eventually die from the accumulation of alcohol and carbon dioxide they produce. The same kind of thing happens in a pond when its plant and animal inhabitants fill it with organic debris and turn it into a meadow, in which aquatic creatures can no longer live. It happens again when a meadow becomes a forest.

Who cares about such episodes? *We all need to.* We need to see that, in each case, the organisms using their habitat unavoidably reduce its capacity to support their kind by what they necessarily do to it in the process of living. In making their habitat less suitable for themselves, organisms sometimes make it more suitable for other species—their successors. This is what mankind has been doing. We have overshot environmental limits and have begun inflicting serious damage upon *our* habitat's capacity to support *our* species.

It may be human to respond to changed circumstances that we do not understand by indulging in puerile blame-placing, or in "retaliating" against supposed perpetrators of our troubles. But it would be more humane to acknowledge that we did not recognize precedents in time to avoid the frustration of ill-founded aspirations.

We can still learn an important lesson by belatedly seeing that the precedents *were* precedents. We can at least recognize that the loss of the American dream is no uniquely human event, so it does not need to fill us with seething hatred, either toward ourselves for some supposed apostasy or malfeasance, or toward some alleged tyrant, villain, or enemy nation.

The rest of this chapter will outline principles that help explain the human predicament. They differ altogether from the rhetoric by which the leaders of nations, parties, industrial organizations, dissident groups, or revolutionary movements purport to account for contemporary history. They are principles we can no longer afford to disregard as "merely academic."

Universal Interdependence

Let us begin with the fact that all the many forms of life on this planet have the same ultimate chemical basis. In many and varied ways, their life processes are also interdependent. Each form (including man) takes certain substances from its environment and puts certain substances into its environment; these affect other organisms. The influences of various species upon each other are often intricate, indirect, and subtle.

One of the first men to gain a really deep appreciation for these complex connections was Charles Darwin, who spoke of the *web* of life. Not only did Darwin painstakingly work out the evolutionary implications of the ways living things interact, but he also called such strong attention to the interconnectedness of the web of life that the subsequent emergence of a science concerned specifically with interactions among diverse organisms was inevitable. That science later acquired the name "ecology"—a name which later came to have too superficial a meaning, for some of the groups among whom it became in the 1970s an "in" word.[2]

Ecology is concerned with far more fundamental ideas than food fads, returnable bottles, or even the banning of aerosols. It is dangerous and unnecessary to remain preoccupied with the peripheral, the particular, or the superficial. The *basic principles* of this science are so important to man's effort to comprehend what is happening to him and to his hopes that they need to be made a part of everyone's common knowledge.

Significant understanding can be developed from the well-known fact that most of the organic world is divided into two "kingdoms"—plants and animals. There is a simple but all-important division of labor between these two kingdoms: they "cooperate" in sustaining life on earth. Green plants, using energy from sunlight, extract carbon dioxide from the atmosphere (or the sea) and use it to form various organic molecules, which other organisms can then use as food. In this process of photosynthesis, plants give off oxygen, which becomes part of the atmosphere. Animals need the oxygen to breathe, so they can oxidize their food and release the solar energy stored in it, just as the solar energy stored in a stick of firewood is released when we burn it. Because animals cannot manufacture the carbon-based organic molecules themselves, they need plants to feed upon (or the flesh of other animals that was formed by eating plants). As animals breathe, they return carbon dioxide to the atmosphere (or the sea). These two chemical elements, carbon and oxygen, circulate in an endless biogeochemical cycle.

This clear interdependence between the plant and animal kingdoms provides a model for understanding more detailed instances of "cooperation" between different kinds of organisms in the web of life.

If there are cooperative interactions, there are also competitive inter-actions. Only man, of course, can apply these somewhat anthropo-morphic but highly illuminating labels to these processes and *under-stand* how such interactions arise, how they work, and how *we* are inescapably affected by them.

When life began on earth, the plant kingdom at first had the place to itself, for animals could not evolve until millions of generations of plants had made oxygen abundant. Originally, there was no free oxy-gen in the earth's atmosphere, just as there appears to be little or none in the atmospheres of the solar system's other planets or the earth-sized satellites of Saturn and Jupiter.[3] The members of the plant king-dom consumed carbon dioxide, which was then abundant in the air and in solution in the sea. The plants gave off oxygen. The cumulative effects of the life processes of the plant kingdom changed the world. (The effects could be cumulative partly because some plant substance from each generation got buried somehow, and the carbon in it was prevented from re-oxidizing. Eventually such buried organic matter was transformed by heat and pressure into coal, etc.; from what has been said in previous chapters, it will be recognized that that was one way these biogeochemical processes acquired political and economic relevance. But even more important, as we shall see in a moment, we humans could not exist if these things hadn't happened.)

As plant life changed the world, forms of plants that could only thrive in an atmosphere rich in CO_2 became extinct when the habitat to which they were adapted had been made non-existent by their use of it. As a further result, other forms of plant life were able to take their place, and animal life also eventually became possible.

As animals evolved, their life processes began to balance the ef-fects of plant life. Animals consumed plants, took in oxygen, oxidized the energy-rich carbon compounds in the plants they ate, and gave off carbon dioxide. After hundreds of millions of years in which many new species emerged and many old species became extinct, an ap-proximate chemical equilibrium was reached between the two king-doms. The rate at which the plant kingdom was changing the world in one direction was now just about offset by the rate at which the animal kingdom was changing it back in the other direction.

That equilibrium was reached long before man appeared. Evolu-tion could not produce any creature with the traits of man until the world had undergone the changes that would enable it to support such a creature.[4] Or, to put it the other way around, we are the kind of creature we are because our species evolved in a habitat that had already been brought to the conditions recently known by our not-so-distant ancestors. However, *like the life processes of every previous species, human life processes modify the environment upon which we depend.* That is a law which no creed can repeal.

As human numbers and equipment increased, human impact on the environment increased, so it departed faster and faster from being the kind of environment in which our species had evolved. For example, in the prehuman environment much of that carbon removed from the atmosphere by green plants was locked safely away in the earth, where it could not be returned to the air by respiration. Disregarding his own need for a nearly carbon-free atmosphere, man perceived the deposits of coal and petroleum not as safe underground storage of natural pollutants, but as "fossil fuels"; he set about eagerly unearthing them to fulfill his growing demand for energy. By energy-releasing combustion he was returning the carbon to the atmosphere at a rate that has now imparted a significant imbalance to the carbon cycle. Experience of submarine crewmen has shown where such imbalance can lead us. For example, in 1973, the chemicals in the air purification system aboard a submerged undersea research craft proved insufficient when the sub became entangled in a ship's wreckage; two members of the research crew shared the plight of the zymogenic micro-organisms in the wine vat, their lives ended by the accumulated autotoxic products of their own life processes.[5]

Photosynthesis and the Basis of Life

If breathable air is essential to us, so is energy. Energy from the sun is the ultimate basis of all life. Solar energy enters the web of life by way of green plants. By the light-powered chemical process called photosynthesis, plants produce all the food for the animal kingdom. Sunlight falling on an animal's body contributes to its warmth, of course, but cannot energize its muscles and other organs. For that the animal must depend on solar energy that was earlier trapped and stored in chemical form in the plants at the beginning of its food chain.

Such dependence is transitive, as the term "food chain" signifies. This important concept embraces the following kinds of relationships. First, as we have seen, most plants manufacture their food from abiotic (inorganic) substances. Some of the organic matter in plants is consumed by animals. In the bodies of herbivores (vegetarian animals), organic molecules of plant origin are converted into flesh and blood, or eggs, or secretions such as milk. Animals may eat other animals' eggs. Carnivorous animals may eat herbivorous flesh; some eat lower-order carnivores. All organisms, in short, are linked to other organisms in a hierarchy of consumption. Organisms thus consist of substance that undergoes a continual flow upward through a food chain.

Animals, then, are consumers; most plants are producers. But certain kinds of plants (as well as animals) are consumers of detritus. They reprocess organic matter that already had accumulated in their habitat—from fallen leaves of living trees, or the log of a dead tree, or from the decomposed bodies of other organisms on the ground or at the bottom of a pond. The implications of a life style based on detritus consumption will be examined in Chapter 10.

Not all organisms occupy fixed positions in a food chain. Some animals, for example, are partly or occasionally herbivorous and partly or occasionally carnivorous. We are of this type; we can shift our position in the food chain upward or downward. With each link between us and the bottom of the chain, roughly 90 percent of the substance and energy from the lower link are lost into the inanimate habitat. Conversion from one living form to another is about 10 percent efficient. The food available for human consumption is more abundant, as the public began to learn in the 1970s, if man places himself low in the food chain. Food is less abundant as we place ourselves higher (by eating more meat, eggs, or dairy products, instead of cereal grains). It is about ten times more difficult to raise the *quality* of the human diet than it is to increase the *quantity* of food available for human consumption.[6] The web of life can include fewer human beings the more carnivorous they insist on being. But man is so constructed that he needs particular nutrients that are not easily included in an all-vegetable diet. Protein deprivation in infancy, for example, can make him less human by seriously impairing brain development.

Symbiosis and Antibiosis

The interrelations between organisms are not always the interrelations of the eater and the eaten. There are many instances of unwitting cooperation (symbiosis) between one kind of organism and another. Herbivorous animals, for example, are dependent upon symbiotically coexisting bacteria within their bodies. The bacteria contribute to the digestive processes of the host animal. The host animal provides the moist, warm, dark conditions required by the bacteria and gathers the food upon which both it and the bacteria subsist. Flowering plants provide food in the form of nectar for flying insects;

the insects provide a service in pollinating the plants and enabling them to reproduce. Legumes such as clover and alfalfa have on their roots nodules of bacteria that are able to capture nitrogen from the air and convert it into substances needed by the host plants. In fact, the symbiosis between the leguminous plant and its associated bacteria can "fix" nitrogen in sufficient quantity so that it enhances the fertility of the soil, preparing it to support other, humanly useful plant life.

Within either the plant or the animal kingdom, different species make different demands on the environment. When the environmental demands of different species are mutually complementary, so that one species puts into the environment what the other needs to take out and vice versa, their relationship is symbiotic.[7] While we have already seen that this kind of relationship holds, in a broad way, between the whole plant kingdom and the whole animal kingdom, here we are taking note of its occurrence between differentiated species within a kingdom. In other words, there is symbiosis in detail, as well as symbiosis in general.

Now let us pursue some of the ramifications of this idea so that we can see how it applies to man and why it is perilous to disregard it. Let us suppose we have just visited a zoo, or a botanical garden in a city park, or have just been doing some weeding in a small window-box. We are therefore acutely aware of the fact that organisms come in many forms. But now let us remind ourselves that what matters ecologically is not the conspicuous fact that two species have different shapes or appearances, but the less obvious fact that they require different things from their environment and do different things to it. Speciation, the separation of organisms into distinct non-interbreeding types, has ecological significance largely as an indication of "occupational" differentiation, rather than differentiation of form.

To this way of understanding the importance of differences between species we must add the fact that, among the higher forms of life, increasingly elaborate symbiotic relationships *within* species emerge. Different members of the *same* species can *behave* differently from each other, making somewhat different demands on the environment. Man in particular is one species that can become so behaviorally differentiated that it is *as if* he were divided into many species. The interactions between human roles, or different labor-force occupations, are functionally equivalent to the interactions between diverse species. Each role has a distinctive configuration of competitive and symbiotic relationships to organisms occupying other roles. Thus many of the ecological principles that describe relationships between species can also be used to understand relationships between functionally differentiated categories within the human species. The enormous capacity of *Homo sapiens* for behavioral differentiation turns many ecological principles into sociological principles.

As was pointed out by Sumner, the sociologist who wrote "Earth Hunger," the ratio of a population to the available quantity of a necessary resource is an important determinant of the intensity of competition. This ecological principle, so clearly applicable to human society, was tragically neglected by most of Sumner's contemporaries. Nevertheless, when each individual in a population affects the other members of that population by using up some of the common supply of a resource upon which they all depend, there is competition among them.

Insofar as the members of a particular species tend to make identical demands on their environment, their relationship to each other tends to be competitive. It will be only mildly competitive as long as population density is low relative to the resources required by the species. As the population becomes more numerous, competition between its members must be intensified.

Not only does every organic species take substances from its habitat in the process of living; it also puts substances into the habitat. The life processes of many organisms put into their habitat certain soluble chemical compounds whose presence affects the life processes of these organisms themselves, and of other organisms sharing the same habitat. Some of these "extrametabolites" act as inhibitors of growth; others act as promoters of growth. The antibiotics used in medical practice are extrametabolites.[8] The word "antibiotic" is merely the opposite of "symbiotic." The penicillium mold, or fungus, produces an extrametabolite that inhibits growth of various bacteria. The bacteria it can inhibit include many that cause disease in the human body. In refined form, therefore, this extrametabolite (penicillin) has wide-ranging medical use. Few of the people whose lives it has prolonged have understood that they were beneficiaries of an ecological process. Likewise, few of those in whose vocabulary "ecology" means merely the pious denunciation of waste or pollution have recognized the full scope of ecological knowledge. Accordingly, the tragic neglect of ecological understanding by political leaders is pardonable.

When the chemical by-products of the life processes of one species are harmful to another species, their relationship to each other is antagonistic. The penicillium mold is antagonistic to various disease-causing bacteria. (From the point of view of the bacteria, penicillin exuded by the mold is a "pollutant" of the environment.) Similarly, the man-and-machine systems of Los Angeles are antagonistic to the ponderosa pines in the San Bernardino National Forest; the smog derived from automobile exhaust and industrial fumes is a kind of extrametabolite exuded by megalopolis and is toxic to other organisms such as trees.

Obvious as it might seem, it is worth noting that, in the ecological context, antagonism is quite impersonal. The word does not have the emotional connotations it would have in reference to human conflict. There is neither animosity nor vituperation between the people of Los Angeles and the ponderosa pines, or between the penicillium mold and the bacteria. This emotionless meaning of the word is noteworthy because this impersonal kind of antagonism can happen between human beings just as truly as it happens to other living things. Antagonism in ecology merely means that the fulfillment of one organism's needs is antithetical to the maintenance of environmental conditions in which another organism's needs can be fulfilled.

Among human beings, however, it may be that ecological antagonism tends to *produce* its emotional counterpart, just as ecological exuberance gave rise to exuberant feelings. Human animosity can perhaps be aroused by mutual interference between human groups, even when the interference is indirect and unintended. If so, that is one way an age of overpopulation could become an age of war. To recognize that human conflict may arise without villainy, from ecological sources, may be an important step toward preventing unnecessary escalation of frictions misinterpreted. Not always blessed are the would-be peacemakers, if they rely too much on exhortation and take no interest in discovering environmental sources of antagonism. If incongruent with the facts of the real world, their hortatory efforts may intensify rather than abate the human frictions they deplore.

Naive liberalism (political and religious) tends to deny the existence and neglect the occurrence of any valid basis for antagonistic relations between human beings. By implication, if not always explicitly, it attributes conflict to mere human perversity. It avoids the question of whether what passes for a character defect has deeper causes—whether or not the ability to practice brotherly love, self-restraint, and a decent respect for the opinions of mankind depends on such environmental prerequisites as low population pressure and relative absence of civilization-made extrametabolites. By ignoring these issues, such liberal ideology probably nurtures its opposites: those ideologies that regard conflict as inherent in the nature of society or the nature of man.

Community, Niche, Succession

Various differentiated organisms which influence each other adapt collectively to the life-supporting conditions of the habitat. A "community" is a more or less self-sufficient and localized web of life making such a collective adaptation. The roles in a community are performed by many different species of living things, both plant and animal, so it is referred to as a "biotic community."[9]

There are biotic communities in which the most significant roles are performed by plants; these are appropriately called "plant communities" even though there may be some animal species playing quite indispensable roles in them—e.g., pollinating insects, seed-dispersing birds, etc. In other circumstances, animals play a more conspicuous part in the collective adaptation of the association of diverse organisms to its habitat; these can be called "animal communities" even though some plant species are indispensably involved. When man, through his own extensive differentiation on social and occupational lines (quasi-speciation), plays so many of the roles that the community appears to be mostly under his control, it is properly regarded as a "human community." But man is not self-sufficient; the human body cannot, for example, make its own food by photosynthesis from abiotic substances. Because of the dependence of the animal kingdom (including *Homo sapiens*) upon the plant kingdom, there can be no *exclusively* human community, nor any exclusively animal community. The phrase "human community," therefore, should always be regarded as shorthand for a biotic community dominated by human beings. Once again, however obvious these points may seem, it is important to remind ourselves that they were not at all obvious in the Age of Exuberance. The illusion of human autonomy has continued to influence national policies right on into the post-exuberant age.

Each kind of organism has its own distinctive functional role in a community's collective adaptation. The distinctive part it plays is that organism's niche. As competition between organisms intensifies, there is a tendency for them to become differentiated in the things they do to the habitat and the demands they make upon it. In short, there is a tendency toward "niche diversification." Each organism in a community competes principally with the limited number of other organisms within its own niche, not with all other organisms in the habitat. Niche diversification is thus an adaptive response to population pressure. It is through this process that evolution has led to greater and greater community complexity. Although the French sociologist Emile Durkheim pointed out the human social significance of such ecological principles in 1893, the ecological nature of his study of *The Division of Labor in Society* has not always been recognized even by sociologists, and his work has not been generally known by a wider public.[10]

Nevertheless, this brings us to the principles most indispensable for understanding the human predicament. A human community, like any other biotic community, is (1) an association of *diverse* organisms (2) *collectively* adapted to (3) the conditions of its habitat. Such conditions are not fixed. Habitats change. Adaptive patterns must change in response. Often a particular association of organisms cannot avoid altering the characteristics of its habitat by its very mode of adaptation to it. *By unavoidably modifying its habitat in the process of living in it, an association of organisms compels itself to change its own mode of adaptation.* In other words, the configuration of niches in a biotic community is not usually static. Community structures change. They change as a result of the community's own impact on its habitat.

Conceivably (but, in practice, rarely) the organisms that are associated in a particular habitat could arrive at an equilibrium, where their collective impact on the habitat leaves it unchanged from year to year. Such a "climax community"[11] can only come about when the assortment of niches within it is such that their environmental effects are mutually complementary, and the population in each niche is kept stable by its interactions with the populations in the other niches. In this equilibrium condition there are no superfluous types of organisms, because none persists without a niche (a part to perform in the collective adaptation). In the climax community, if we "consider the lilies of the field," we find there *is* some way in which they do their share of "toiling" and "spinning."

The climax community is, in short, an integrated and self-perpetuating community. It comprises a combination of species that can successfully out-compete any alternative combination that might otherwise exist in its place.

What does it mean to say it is "integrated" and "self-perpetuating"? Consider again the model of cooperation implicit in the circulation of carbon between the two big kingdoms, plants and animals. In a climax community, the organic fixation of carbon by photosynthesis, or what ecologists mean by "production," is in balance with the return of oxidized carbon to the atmosphere by "respiration." *This is an important point, but it has been disregarded by industrial civilization.* Most communities are not climax communities. They undergo continual change wherein one species is progressively replaced by another. Even more emphatically, most *human* communities (at least in the

105

modern world) cannot be climax communities. Certainly in the modern city photosynthetic production does not match oxidation of carbon (by respiration plus combustion).[12] As the next several chapters will show, man's efforts to do the very things for which his species has special aptitudes have the inescapable effect of fostering this imbalance, so *human ascendancy undermines itself.*

This happens to various associations of non-human species, too. Community transformation results from modification of the habitat by the community that exists in it at a given time. As the habitat changes, the association of plants and animals it will support must change. This is "succession," the process of change from one community type to another.

Sometimes, even without alteration of the environment, a more effectively adapted species out-competes (and replaces) a less effectively adapted species. This too may be called succession, but it is a milder kind. Sociologists have meant this mild kind of replacement process when they have applied the word "succession" to events observed in human communities. The displacement of one ethnic group by another in a given neighborhood of a city has typified what succession means to sociologists. Important as the social repercussions of ethnic turnover may be, and traumatic as the experience may be for some of the individuals involved in the process, this is hardly the most important form of community transformation affecting human beings. As we shall see in the next chapter, because sociologists have paid too little attention to the larger kind of succession whereby a community causes its own demise by what it does to its habitat in the process of using it, their studies have shed less light than they should have on the more serious dimensions of the human predicament.

The entire sequence of community types characteristic of a given site is what ecologists call a "sere," and the developmental steps in the process of community succession are "seral stages."[13] The key idea involving this cluster of concepts is that most biotic communities *are* subject to change because they *do* alter the characteristics of their own environments. Succession is a very common (and virtually inescapable) ecological process. It happens to human communities as well as to animal and plant communities. The Age of Exuberance in which the American dream unfolded was an early seral stage in the succession of New World community types. The post-exuberant age is a later stage in the same sere.

By not recognizing the ecological pattern, we have misconstrued our own history.[14] Our misunderstanding enabled us to overshoot carrying capacity. If misunderstanding persists, it can turn the ecological kind of antagonism into the emotional kind, making an unkind fate crueler than necessary. As other mammalian species have moved into a post-exuberant stage, increased antagonism and competition typically have led to increased violence and behavioral degeneracy. Status hierarchies become more abrasive. Care and training of the young become inept and even reluctant; the young come to be treated as intruders. One result of these changes is a slowing down of the rate of population increase. Another is the suffusion of fear, hostility, and misery into all aspects of life. Such responses to overpopulation of a limited habitat have been observed in careful studies of other primates (baboons, monkeys), as well as in laboratory rats.[15]

Man has imagined himself to be more unlike other mammals than he really is, so when human behavior has shown these same characteristics, various other explanations have been put forth which have obscured the significance of population pressure itself. In the twentieth century, with human numbers enlarged and resource drawdown becoming significant, man went to war. He rioted in the streets. He committed more and more crimes of violence. His political attitudes polarized and he created totalitarian governments, some of which gave license to sadistic tendencies. A generation gap widened and deepened. In spite of earnest efforts by humane activists to inhibit racism and to rectify economic inequality, disparities between people remained and animosities became more virulent. Standards of decency in behavior toward others and expectations of considerate self-restraint were eroded and degraded in many places.

Man's One World

We need to remember that mankind is part of the animal kingdom. The human species is as dependent as the rest of the animal kingdom upon the plant kingdom. As human numbers have increased, an increasing fraction of the plant kingdom's total productivity has been diverted from feeding other animals to feeding man or the animals man uses. One ecologist has estimated this fraction as one-eighth of the net production of the world's land areas.[16] Thus, with only three more doublings of human numbers, we and our domestic animals would be consuming everything else that grows on all the continents and all the islands of the world, and eating it all just as fast as it could be grown and harvested. How many heads of government or members of legislative bodies have begun to face up to the implications of these numbers?

Since man began to shift from hunting and gathering to agriculture some 10,000 years ago, our species has appreciably altered the structure of the worldwide web of life. Human effort has tremendously increased the fraction of that web that consists of human flesh, as we saw in Table 1. In only about 400 human generations—a short time in an evolutionary perspective—the human population has doubled nine times.[17] Nine seems like a small number. But nine *doublings* (that is, increase by a factor of two raised to the ninth power) is a more than five-hundred-fold increase. Since the dawn of agriculture, the world's people have become about 500 times more numerous.

Worldwide, therefore, the effort to divert to human use still larger fractions of the annual product of photosynthesis has become more and more imperative. This task had to become more and more difficult, since the least difficult diversions would obviously have been, in general, the first achieved. Increased difficulty of diverting still more productivity to human use appears to have increased man's nostalgia for earlier and easier times, although the causes of such feelings remained obscure, as long as these ecological relationships had not yet become folk knowledge.

From the early nineteenth century onward, new tools and new techniques gave man increased power to out-compete other members of the animal kingdom in consuming the products of the plant kingdom's life processes. In other words, even though it was becoming more and more difficult to enlarge human carrying capacity by further takeovers, technological advances had further enhanced our power to do so despite the difficulty. We pressed on with the old takeover method; at the same time, we were exuberantly forging ahead with the newer drawdown method. Accordingly, human numbers increased more rapidly than ever—two of those nine doublings during the hundred centuries since the dawn of agriculture were confined to the most recent one and a half centuries. After *mechanized* agriculture began to be the dominant mode of sustenance production, the web of life was altered more extensively and more swiftly than ever before. If, somehow, man's agricultural modification of the web of life (with steel plows, mechanical harvesters, fossil fuel–burning tractors, and synthetic fertilizers and pesticides) were to falter, and the web were to revert to something like the structure it had only six human generations ago, then (because of these two most recent doublings) *four earths* would be needed to support the present human population of this one earth.

With four earthfuls of people (by pre-mechanized standards), and with reasons to doubt the stability of an ecosystem modified by the drawdown method, the post-exuberant world found itself facing serious trouble. The drawdown method of overshooting human carrying capacity had given increased impetus to the older takeover method of extending it. As the human component of the animal kingdom increased, the nonhuman parts of it inevitably had to decrease. The total sustenance base for animals provided by the plant kingdom was not enlarged by man's efforts; recent evidence suggests it has actually been significantly diminished.[18] What human beings did through selective breeding and intensive cultivation of usable plants was to *displace* types we do not consume with types we do consume, or with plants consumed by animals we consume. The illusion that man was making the world "more productive" arose from simply making a larger share of its limited productivity serve human wants. Local extensions of total photosynthetic productivity through irrigation of former deserts were offset by creation of deserts through deforestation, overgrazing, etc., and through turning formerly productive land into deserts of another sort—burying arable land under buildings, highways, mine tailings, junkyards. Industrial man also reduced the life-sustaining capacity of the earth's air and water, diminishing the marine ghost acreage upon which he sometimes assumed he could fall back.

Homo sapiens was becoming a global "dominant." A dominant is a species with greater influence than any other in its biotic community upon the characteristics of its habitat and upon the lives of the other species associated with it.[19] Man apparently regarded himself as a dominant species long before he had paved, poisoned, or plundered vast portions of the planet that produced him. He was conscious of his own dominance even in biblical times, when his "dominion over every living thing that moves upon the earth" was assumed to have been divinely established. But most men have never fully recognized the implications of such dominance for their own future. There has never been a law of nature to guarantee that our species would retain its dominance forever, once we had achieved it. It could almost be said that the laws of nature make it impossible for the dominance of any species to be permanent.[20] Against these considerations, the petulant strivings of nations for political, economic, or military hegemony seem almost grotesquely transient.

The former relative isolation of various parts of the globe from each other meant that the world was effectively divided into many more or less independent biotic communities. Man, however, became a very mobile species. People proceeded to shift other animals and plants from the sites where they first evolved to other habitats where they entered new biotic associations, so much so that we have virtually converted the world into a single biotic community. Human actions now dominate that global community more than ever, for our technological equipment has grown more extensive and more potent (in ways that will be assessed in Chapters 9 and 10).

Dominance, however, is not omnipotence, even for man. The life of any organism depends upon an adaptive fit between its traits and the characteristics of its habitat. The organism's traits are what they are because of evolutionary selection in a habitat that developed before the organism did. But all species produce changes in the habitat that supports them, and these may or may not be offset by other changes produced by other coexisting species. Unless the influences of the various species in a community balance each other, the dominant species will undo its own dominance by its environmental impact.

As the form of a biotic community changes through succession, one dominant species replaces another. Succession thus involves several alternative fates for the dominant species: loss of dominance, extinction, or migration elsewhere. If the species cannot relocate, but the conditions that enabled it to thrive and attain dominance have changed, it *may* continue to live as a subdominant member of the community. If it cannot do that, it may cease to live in its former locality. If the species is mobile, it may migrate elsewhere to avoid loss of dominance or extinction.

For modern man there has ceased to be any "elsewhere." History has seen *Homo sapiens* preserve his dominance more than once by migrating from one local habitat that he had changed to another that he proceeded to modify from the moment of his arrival. Man has migrated to a new "promised land" time and time again, after exhausting the old country's soil, or fishing out its lakes and streams, or killing off too much of its game, or harvesting the timber that once made it rich, or playing out its mineral desposits, or pumping dry its oil fields. The "ghost towns" dotting the American West testify to the relevance of succession in human history.

John Steinbeck's 1939 novel, *The Grapes of Wrath,* depicted the plight of thousands of people who migrated from ruined Oklahoma farmlands to California—where other people were already dominant and were reluctant to share their precarious niche with new arrivals. Earlier, during the sixteenth, seventeenth, eighteenth, and nineteenth centuries, overpopulated Europe spilled scores of millions of its excess people into the New World. The people already inhabiting the Western hemisphere, the Indians, were neither sufficiently numerous nor technologically equipped to resist the European invasion effectively; nor could they have dominated and appreciably changed the diverse biotic communities upon which their various tribes had depended. The Europeans brought with them the power to change everything—everything except the laws of nature. They could not change the principles of ecological succession which would ensure that their dominance was less than eternal.

Four hundred years of European (and American) dominance of the global community had changed so much of the world that less and less of it resembled the habitat in which *Homo sapiens* had evolved and to which he was genetically fitted. The once separate human communities scattered over the earth coalesced under the impact of European-American technology and culture into a single but inharmonious community. This came about in an age when the dominant people happened to be Europeans and then Americans. But no law of nature required that the dominant position must forever be reserved for Europeans or Americans. On the contrary, natural precedents strongly suggest that man's dominance in the global biotic community is temporary. Likewise, the dominance of Americans in a global human community had to be temporary. It was temporary for the British, upon whose empire the sun finally did set—in the years following an ostensibly successful expenditure of blood, sweat, and tears to defend the land of hope and glory from loss of dominance to the Luftwaffe and the Wehrmacht.

New Attitude Needed

Contemporary history continues to be badly misunderstood unless we have in our working vocabulary these clusters of concepts:

web of life	competition
photosynthesis	niche diversification
food chain	speciation, quasi-speciation
producers, consumers	biotic community
biogeochemical cycles	dominance
symbiosis, antibiosis	succession, seral stages
extrametabolites	climax community
antagonism	detritus ecosystem

A highly unfamiliar attitude toward change is required and becomes possible if we add these ecological concepts to our normally myopic economic-political vocabularies. Then we can begin to apply principles containing these concepts to the human situation.

111

In nature, community succession (and the displacement of one dominant species by another) results from habitat transformation, not from violent conquest. It cannot be prevented by any actions resembling military defense. A dominant species cannot fight off its prospective successor; it may or may not surpass it in successfully adapting to environmental conditions.

A natural sequence of succession took us from the Age of Exuberance and the American dream into the age of overpopulation and the global trauma. Man's war against his insect competitors and his bacterial predators was, it seemed, a resounding success. However, its sequel is going to be as unsought and unwelcome as the sequel to Anglo-American success in the war Winston Churchill thought need not, if won, lead to "dissolution of the empire." Judged in retrospect by its consequences, the mechanization of agriculture may seem more of a curse than a blessing. The so-called Green Revolution, welcomed with the 1970 Nobel Peace Prize for development of the high-yield grains that fuel it, may burden the close of this century with almost another doubling of world population.[21] The result will probably be further intensification of man's substitution of fratricidal behavior for fraternal behavior. The American dream may be made twice as irretrievable as it is already.

It was a fact of life that frontier America (with its partially achieved promise of liberty, equality, and progress) was not, and could not be, a climax community. No rewriting of history on ideological lines could change that. Nor can a fossil-fueled industrial civilization prevail as a climax, affording affluence forever. Disease, starvation, or another war may possibly decimate mankind, in which case another brief age of exuberant recovery for the survivors might follow—until the losses had been replaced. That way of reverting to the earlier seral stage whose passage we now mourn can hardly be welcome.

Hindsight tells us it would have been vastly better had the human propulation had the foresight to stop its own increase two or more doublings ago.

Notes

1. The fact that the genocidal response was not unique to Nazi Germany will be further developed in Ch. 13.
2. An extreme instance of the superficial conceptualization of "ecology" is given by Worthington in Rogers 1972, p. 7.

3. See Brancazio and Cameron 1964 (among references listed for Ch. 3).

4. Cloud 1978, p. 117; Fogg 1972, pp. 96–100.

5. Reported in articles in the London *Times,* June 20, 1973, p. 1; June 21, p. 8.

6. It is now known, of course, that there can also be *too much* food of animal origin in our diet; the incidence of both heart disease and cancer tends to be increased thereby. It remains generally true, however, that quality differences in national diets can be roughly assessed in terms of "percent food of animal origin"—if not "the higher the better," then at least "the less the poorer."

7. More precisely, this aspect of symbiosis (or "living together") is called *mutualism,* a strong interdependence between different but associated life forms. See Odum 1975, pp. 140–144.

8. Ibid., pp. 155–157.

9. Cf. Odum 1971, pp. 140ff.

10. Durkheim (1933, pp. 266–270) clearly recognized the sociological importance of what Boughey (1975, p. 17, listed among references for Ch. 2) calls "resource partitioning."

11. Odum 1975, p. 152.

12. Odum 1971, p. 98; Bolin 1977.

13. Odum 1971, p. 251.

14. The depth of our misreading of history becomes apparent when significant books seeking to correct it are examined. See, for example: Carson 1962; Day and Day 1964; Graham 1970; Dasmann 1972.

15. See Russell and Russell 1968 (listed among references for Ch. 13), and John B. Calhoun, "Population Density and Social Pathology," *Scientific American* 206 (Feb., 1962): 139–146.

16. Odum 1971, p. 55.

17. See Table 1, in Ch. 2.

18. See comments by Odum (1971, pp. 43, 46, 48) regarding differences between gross and net productivity, and nature's tendency to maximize the former as contrasted to mankind's effort to maximize the latter.

19. See Elton 1927, pp. 9–10, 13–15, 29; Krebs 1972, pp. 540–543; Odum 1971, pp. 143–145.

20. An eminent British paleontologist, H. L. Hawkins, pointed out more than four decades ago that although mankind was able "to become the most successful type of animal that has ever existed . . . the reward of success in that direction is death." Quoted in Raymond Pearl, *The Natural History of Population* (New York: Oxford University Press, 1939), p. 288.

21. Cf. the comments on implications of "miracle" rice and wheat in a context of "energy subsidies," in Odum 1975, pp. 17–21; see also Michael Allaby, "Green Revolution: Social Boomerang," *The Ecologist* 1 (Sept., 1970): 18–21.

Selected References

Beck, William S.
1961. *Modern Science and the Nature of Life*. Garden City, N.Y.: Doubleday/ Anchor Books.
Bolin, Bert
1977. "Changes of Land Biota and Their Importance for the Carbon Cycle." *Science* 196 (May 6): 613–615.
Carson, Rachel
1962. *Silent Spring*. Boston: Houghton Mifflin.
Cloud, Preston
1978. *Cosmos, Earth, and Man*. New Haven: Yale University Press.
Dasmann, Raymond F.
1972. *Planet in Peril?* Harmondsworth, Middlesex: Penguin Books.
Day, Lincoln H., and Alice Taylor Day
1964. *Too Many Americans*. Boston: Houghton Mifflin.
Durkheim, Emile
1933. *The Division of Labor in Society* Trans. George Simpson. New York: Macmillan.
Elton, Charles
1927. *Animal Ecology*. New York: Macmillan.
Fogg, G. E.
1972. *Photosynthesis*. 2nd ed. New York: American Elsevier.
Graham, Frank, Jr.
1970. *Since Silent Spring*. Boston: Houghton Mifflin.
Hawley, Amos H.
1950. *Human Ecology*. New York: Ronald Press.
Krebs, Charles J.
1972. *Ecology: The Experimental Analysis of Distribution and Abundance*. New York: Harper & Row.
Odum, Eugene P.
1971. *Fundamentals of Ecology*. 3rd ed. Philadelphia: W. B. Saunders.
1975. *Ecology: The Link Between the Natural and the Social Sciences*. 2nd ed. New York: Holt, Rinehart and Winston.
Rogers, Paul, ed.
1972. *The Education of Human Ecologists: Proceedings of a Symposium Held at The Polytechnic, Huddersfield, 28–29 March*. London: Charles Knight.

7 | Succession and Restoration

From Historic to Nondescript

New devices have made people more and more successful in their competition with other species, but often at the cost of obliterating visible reminders of their heritage. Thus, after a fifteen-year absence from the town he loved, the former rector of Bruton Parish found changes that appalled him when he returned. The quiet town of Williamsburg had seen change before, but when its principal link to the rest of Virginia was a dusty carriage road, such change came slowly. During his previous residence there, the Reverend Dr. W. A. R. Goodwin had thus been privileged to walk in the imagined presence of great eighteenth-century Virginians for whom Williamsburg's streets and many of its surviving buildings had been the scene of their formative political endeavors.[1]

Before he left in 1908, Dr. Goodwin had spent six gratifying years as pastor of the old brick church on Duke of Gloucester Street. In 1923 he was returning to take up a position at the College of William and Mary. Americans had meanwhile participated in a war in Europe, worldviews had been restructured, and at home the automobile had taken hold as a major factor in an emerging new way of life. In a natural but regrettable process of local succession, colonial houses and insufficiently appreciated historic buildings were giving way to gasoline stations, garages, and new stores—in which one could no longer feel the presence of George Washington, Thomas Jefferson, George Mason, or Patrick Henry. Utility poles now disfigured once tree-lined streets, obscuring visions of the politically monumental events that occurred there six generations before. Williamsburg seemed to be sacrificing its heritage in a thoughtless quest to become just another modern town.

In Dr. Goodwin's eyes, Williamsburg was special. He did not have to be against change and modernism elsewhere to regret the change and modernization he saw happening in this particular town. Ven-

115

erable buildings, streets, and landscapes were being replaced by new and useful but historically nondescript successors. He may not have known how irresistible succession often is in biotic communities, but in this particular community he knew great efforts must somehow be made to resist unwelcome trends.

Efforts to turn back tides are often fraught with futility, but Dr. Goodwin's effort was extraordinary. It arose from passionate conviction that an older Williamsburg had symbolized a worthy heritage. This made him feel it was in the interest of mankind for *that* Williamsburg to remain visible.

Past Succession

The past Dr. Goodwin hoped to preserve had not been Williamsburg's earliest seral stage. Growth and succession were not just twentieth-century phenomena; many of the historic events for which the town was known had occurred when it was into its second century, already much changed from its first decades.[2] Some of the earliest changes experienced by Williamsburg, and especially the succession of uses to which one of its chief building sites had been put, will serve to exemplify natural and non-extraordinary processes characteristic of all kinds of biotic communities. They have been well studied in plant and animal communities.[3] These changes in Williamsburg will also show how truly the mission this earnest man conceived for himself was one that involved the arrest (and, by grace of generous philanthropy, even the reversal) of a sere.

English settlers had first established themselves on the North American continent at nearby Jamestown. One generation later, that hardwon beachhead of colonization had enabled other English settlements to come into existence in the vicinity. One of them, slightly farther inland, was called Middle Plantation. It remained for its first sixty years a rather insignificant village, until it and the Virginia Colony had grown populous enough to support a college. When William and Mary, the second institution of higher learning (after Harvard) to be established in the English colonies, was founded, the town became an intellectual center, and it was soon apparent that it should therefore also be the colony's capital. Government was transferred there from Jamestown in 1699, and the town of Middle Plantation was renamed Williamsburg. It remained the seat of government until continuing growth of the colony and further shifts of population inland led to selection of Richmond, forty miles farther up the York Peninsula, as a more centrally located capital in 1780.

Seven years after the colonial government moved to Williamsburg, the General Assembly appropriated £3,000 for construction of a Governor's House. The building was not completed until 1720, by which time it had come to be called the "Palace." As one royal governor followed another, the house was enlarged and extensively remodeled, becoming quite elegant with the addition, for example, of a handsome ballroom wing. It was surrounded by magnificently landscaped grounds and a number of ancillary buildings.

When men build a building, they are modifying for their own use a localized bit of habitat. When they remodel or enlarge that building, they are further modifying their earlier modification. Their propensity to do such things is not extraordinary; it is an adaptive trait they share with beavers, birds, wasps, moles, prairie dogs, etc. There was nothing extraordinary about the fact that the Williamsburg Palace, a building intended for official occupancy, grew as the importance of the office grew (as the population of the colony grew). Nor was there anything extraordinary about the fact that the house was eventually destroyed; such mishaps happen. After its destruction, reuse of the remaining foundation, out-buildings, debris, and grounds was to be expected.

Early in the morning hours of June 8, 1775, the last royal governor of Virginia left this Palace with his family and several servants, taking refuge aboard a British man-of-war in the York River. The mansion remained vacant for a time, in possession of the General Assembly and presumably cared for by slaves who remained as part of the property. But in 1776 the Palace Green served as a parade ground for Revolutionary troops, the governor's park was occupied by cavalry, and the general commanding the troops used the Palace briefly as his residence. When Virginia proclaimed itself a commonwealth and chose Patrick Henry as its governor, the Assembly appropriated £1,000 to refurnish the Palace for his use. Thomas Jefferson succeeded him as governor three years later and lived in the Palace briefly, until moving to Richmond.

Again the building was vacant for a while. When Washington's army besieged Yorktown, only a few miles away, the Palace became a military hospital. Among the wounded soldiers of the Revolutionary army who were brought to it, more than 150 died and were buried in unmarked graves in its gardens. In December, 1781, a three-hour fire burned the Palace to the ground. Arson was suspected. The following year an agent of the governor was authorized to sell the reusable bricks. Two buildings flanking what had been the front entrance to the Palace were also sold. (One had been the governor's office.) They were now used as private residences, and remained standing until the middle of the Civil War.

After the Civil War a school was built on the land just in front of the Palace, with its rear wall based on the Palace's front foundation. Presumably the school was built in that precise location because the Palace provided a usable foundation. Moreover, now that the more industrialized North had defeated the plantation South, the states of the secessionist Confederacy would enter the industrial age (following the rest of the Union they were obliged to rejoin). Thus, some years after the Civil War, a factory was put up at the north end of the old Palace garden, and a railroad ran across the Palace grounds. After additional increase in the population and further change in the patterns of American life, a large high school was erected just at the head of the Palace green, the erstwhile parade ground of Revolutionary soldiers.

Thomas Jefferson had helped give impetus to education in Virginia; it would not have suprised him to revisit Williamsburg and find new school buildings. He was also keenly interested in science and technology, so despite his preference for an agrarian America he might not have been astonished by the factory, or even by the railroad. Onset of industrialization in America, and eventually in this small Virginia community, was a natural consequence of economic forces unleashed by expansion of European population and culture into the New World.

Turning the Tide

In retrospect, these forces and their effects were not difficult to understand. But neither were Dr. Goodwin's efforts to retrieve from the twentieth century an eighteenth-century version of Williamsburg.

Some of the successive uses of this one site became known from the meticulous archeological research that was undertaken to implement Dr. Goodwin's tide-turning dream.[4] Maps in possession of the College of William and Mary showed the arrangement of buildings in colonial times. The remains of 156 Revolutionary War soldiers who had died in the Palace were found buried in straight rows in the former garden when the site was excavated to disclose well-preserved foundations. A floor plan of the Palace, drawn by Thomas Jefferson, was obtained from the Massachusetts Historical Society. A copper engraving that showed how the front of the building appeared between 1732 and 1747 was found in Oxford, England. Journals of the House of Burgesses contained informative references to the Palace, and inventories of its furnishings had been made by three colonial governors.

From such information it became possible to undo changes that had occurred since colonial times. Newer buildings were removed and the Palace was reconstructed on the original foundations. All told, in Williamsburg, more than 600 modern buildings were removed to make way for reconstruction of scores of their colonial predecessors. Some 80 colonial buildings were still standing when the work was started, and they were restored. More than 60 colonial gardens were reestablished.[5]

For our purposes, the reestablishment of those gardens takes on particular importance. It helps to highlight important features of the restoration process, and to show just what *succession* really means.

The reestablishment of Colonial Williamsburg itself was ecologically equivalent to the restoration of a garden. A garden is a *contrived* ecosystem; it requires continuous attendance by a gardener to keep it in its desired state. His efforts must be directed toward compensating for the soil changes caused by its occupants, and toward preventing the invasion of unwanted plants (weeds). When a garden has long been unattended and has advanced into a later seral stage, reestablishing it as a garden means artifically undoing the natural succession. Similarly, restoration of the colonial version of the town meant artificially undoing a human instance of succession. This human succession had occurred as naturally as the plant succession in the gardens.

To restore a garden, successor plants must be removed, soils must be restored to earlier conditions, and earlier occupants of the site must be reintroduced. To restore an earlier seral stage in Williamsburg itself, it was necessary similarly to remove some 11,000 lineal feet of concrete sidewalks and replace them with such earlier materials as oyster shells, brick, flagstone, and gravel. Fences, benches, garden walls, hitching posts, and halyard signs for shops and taverns had disappeared as Williamsburg moved through the nineteenth century into the twentieth; these had to be replaced, as did the old streets and buildings. Arresting succession and undoing it was an expensive, difficult, and time-consuming business.

The Raleigh Tavern (a building that served for at least 117 years as a social and commercial gathering place for Williamsburg and all of Virginia, and as a center for seditious secret caucuses just before the Revolution) was the first colonial building to be reconstructed on original foundations and opened to the public. From colonial tavern-keepers' inventories the new Raleigh Tavern was refurnished with furniture and accessories authentic to the colonial period. It was here, in the original building, that the first chapter of the Phi Beta Kappa Society was said to have been formed in 1776 by students at William and Mary.

In 1924 a New York banquet of Phi Beta Kappa was addressed by Dr. Goodwin, and was attended by the extraordinarily wealthy John D. Rockefeller, Jr. Contacts between these two men continued; more than two years later, when they conversed privately at a formal Phi Beta Kappa dinner at the college in Williamsburg, Mr. Rockefeller decided to finance the restoration of Colonial Williamsburg. It took more than $70,000,000 of Rockefeller money to accomplish this undoing of the effects of natural succession in one small but historic human community.

The Struggle against Succession

So large an expenditure just to restore a portion of one small town to a former seral stage surely indicates that succession is a powerful process in human communities. Imagine how costly such restoration would be on a scale required by a community such as New York, Boston, or Philadelphia! The world's wealthiest philanthropist could hardly begin to finance an anti-succession project of such magnitude.

Community restoration efforts on the seemingly monumental scale exhibited in this one small town in Virginia were a minute sample of something *Homo sapiens* had been doing ever since the dawn of agriculture. In particular, reestablishment of Williamsburg's colonial gardens by removal of successor plants, restoration of soils to an earlier condition, and reintroduction of earlier plant occupants of the site represented, in miniature, a process upon which mankind has become increasingly dependent over the last 10,000 years. What the Williamsburg example reveals is this: if human numbers have been able to rise from a few million to several thousand million in only ten millennia (as we saw in Chapter 2), it was because *human beings had learned how to devote their efforts to impeding succession.* A farm (no less than a garden or a town) is a contrived ecosystem. It is a greatly simplified plant community. In successional terms it is thus a very early seral stage, and is therefore inherently unstable. Abandoned, a farm will undergo natural succession toward a most unfarmlike climax, some kind of wilderness.[6] Agriculture, ecologically understood, is the continual undoing of succession.

As we noted at the beginning of Chapter 5, the ruthless displacement of unwanted plants and animals was part of the takeover process by which American pioneers inflicted European agriculture upon the New World. Nature's tendency is to turn farms into wilderness; man's tendency during the past 10,000 years has been to turn wilderness into farms. The Europeans who came to the New World didn't merely fight Indians; they also fought succession, thereby claiming more of the New World's carrying capacity for *Homo sapiens* than the Indians had done.

All humans alive today, being so many times more numerous than their pre-agricultural predecessors, depend for their existence upon man's success—thus far—in artificially converting a climax community (with minimal but permanent human carrying capacity) into a less mature seral stage (with greatly enlarged but precarious human carrying capacity). Although a climax community, or wilderness, has more inherent stability than a farm, it won't feed as many human beings—because more of its total net production is consumed by other species.

As we have seen, the drawdown method augments human carrying capacity on a strictly temporary basis. We have assumed so far that the takeover method yields *permanent* increments of carrying capacity. Now, relating this method to succession, we can see that the gains are precarious. The increments are only *potentially* permanent.

Agriculture was much more than simply a convenient means of obviating man's need for good luck in hunting and foraging. It did more than simply replace uncertainty with routine. It was like a dike or a levee, requiring constant maintenance and perpetual improvement, to hold back the pounding floodwaters of succession (and mass starvation). In a populous, post-exuberant world, an ecologically rational government would not imagine it had created a "Department of Defense" for its people by combining its former War and Navy Departments and heavily budgeting the combination. Instead, to defend its people from the *real* enemies to their security, it would combine its Department of Agriculture, its Public Health Service, and its National Science Foundation, and would give the lion's share of its budget to *that* combination.

Succession Misunderstood

Continually increasing world population became more and more dependent on this continual struggle to undo succession. But the culture of exuberance made almost everyone (including for a while even those sociologists who were beginning to call themselves human ecologists) blind to this important conception.[7] We insisted there was, as Roderick D. McKenzie said, "a basic difference between human ecology and the ecologies of lower organisms." That difference, we believed, lay in man's capability of what McKenzie called "a higher level of behavior" by which, as a cultural animal, man "creates . . . his own habitat."[8] The culture of exuberance seemed to impute almost supernatural capabilities to *Homo sapiens*. It prevented us from seeing that the process of "creating our own habitat" might be a trap, that technology might come to enlarge our resource appetites instead of our world's carrying capacity.

An important volume on *Plant Succession* was published in 1916 by Frederic Clements. It recognized the fact that the kind of biotic community that can thrive in a given habitat depends on the characteristics of that environment and how those characteristics act upon the organisms in the community. It also clearly recognized that the organisms, in the process of living, *react* upon the habitat.[9] Plants change the soil in which they grow. Tall plants provide shade, making the site suitable for shade-tolerant species that could not have grown there before. Plants that require stable conditions of moisture may thrive only after other plants have established a moisture-holding ground cover. The reaction of a community upon its habitat, Clements saw, was at the heart of the process of succession. Early seral stages "pave the way" for their own replacement by later seral stages.

Almost immediately, a few sociologists (such as Robert E. Park, of the University of Chicago) began to apply some of the ideas from Clements and other ecologists in their efforts to understand human experience.[10] While they talked of succession, most of them missed this key insight into its nature. Students of Park's colleague, Ernest W. Burgess, came to regard succession in human communities essentially as a process of aggression; invaders were imagined to be succession's driving force, pushing out prior occupants who supposedly might otherwise have thrived forever on a given site. The fact that *occupants of a site may make it unsuitable for themselves* after a time, by the use they have made of it, was not recognized. This idea was incompatible with the culture of exuberance, so the sociologists who pioneered a specialty they began calling "human ecology" did not grasp it.

The idea that man's dominance of a world ecosystem might be only a pre-climax stage in a sere with other stages to come (not dominated by man) was quite alien to the culture of exuberance. The idea that industrial man's impact upon his habitat might make it unsuitable for industrial man clashed with the prevalent idea that human control over nature was a great achievement in the exploitation of limitless resources. But only ten years after McKenzie's intimation of human exemption from ecological principles of a kind that applied to lower organisms, Amos Hawley insisted that "the only conceivable justification for a human ecology must derive from the intrinsic utility of ecological theory as such."[11] Hawley further affirmed in his later writing that man was like every other living creature in being "inextricably involved in the web of life." All life requires the association of diverse species, he said. Human wants had multiplied, and so had the techniques for their satisfaction. Rather than exempting our species from ecological principles, according to Hawley this had "served to implicate man more thoroughly than ever in the natural environment." *Homo sapiens*, wherever found, was "an integral part of a biotic community."[12]

By the 1960s other sociologists, such as Otis Dudley Duncan, began to adopt this perspective and narrow the gap between human ecology and general ecology.[13] But going into the 1970s their insights were still largely eclipsed by public beliefs left over from the culture of exuberance. By this time, human self-displacement by the reaction of industrial civilization upon its habitat had proceeded much further than when the Williamsburg restoration began. The dependence of mankind on complex technology relying on ghost acreage had burgeoned, and the earth's human load had more than doubled in the half-century since Robert Park first saw the human relevance of Clements's studies of plant succession.

Notes

1. The story of Dr. Goodwin's return to Williamsburg and the consequences thereof is nicely told in Chorley 1953.
2. See Lewis and Walklet 1966.
3. See various papers included in Golley 1977.
4. Visitors to the restored historic town are introduced to examples of the

careful research underlying the restoration by exhibits at the Information Center and the Archaeological Museum. See Colonial Williamsburg 1964, pp. xviii-xix, 1–2, 40–41, 89.

5. Ibid., p. xvii.
6. See, in particular, the papers in Golley 1977, by Crafton and Wells (pp. 55–76) and Odum (pp. 102–117).
7. For a notable exception see Mukerjee 1932. Probably his familiarity with situations in India, where man's membership in (rather than outright dominance over) biotic communities was more visible, sensitized him to ecological ideas whose sociological relevance would remain obscure for another generation or two of American sociologists.
8. Hawley 1968, p. 41.
9. Clements 1916, pp. 79–110; note especially pp. 80–81.
10. See the adaptation from Clements in Robert E. Park and Ernest W. Burgess, *Introduction to the Science of Sociology* (Chicago: University of Chicago Press, 1921), pp. 525–527. See also Park 1936.
11. Hawley 1944, p. 399.
12. Hawley 1950 (listed among references for Ch. 6), p. 55.
13. Duncan 1961; 1964.

Selected References

Chorley, Kenneth
1953. *Williamsburg in Virginia: Proud Citadel of Colonial Culture.* New York: Newcommen Society of North America.
Clements, Frederic E.
1916. *Plant Succession: An Analysis of the Development of Vegetation.* Washington: Carnegie Institution.
Colonial Williamsburg
1964. *Official Guidebook and Map.* Williamsburg, Va.
Duncan, Otis Dudley
1961. "From Social System to Ecosystem." *Sociological Inquiry* 31 (Spring): 140–149.
1964. "Social Organization and the Ecosystem." Ch. 2 in R. E. L. Faris, ed., *Handbook of Modern Sociology.* Chicago: Rand-McNally.
Golley, Frank B., ed.
1977. *Ecological Succession.* Stroudsburg, Pa.: Dowden, Hutchinson & Ross.
Hawley, Amos H.
1944. "Ecology and Human Ecology." *Social Forces* 22 (May): 398–405.
1968. (Ed.) *Roderick D. McKenzie on Human Ecology.* Chicago: University of Chicago Press.
Lewis, Taylor Biggs, Jr., and John J. Walklet, Jr.
1966. *A Window on Williamsburg.* New York: Holt, Rinehart and Winston.

Mukerjee, Radhakamal
1932. "The Concepts of Distribution and Succession in Social Ecology." *Social Forces* 11 (Oct.): 1–7.
Park, Robert E.
1936. "Succession, An Ecological Concept." *American Sociological Review* 1 (Apr.): 171–179.

8 Ecological Causes of Unwelcome Change

Biotic Potential versus Carrying Capacity

In the next seven generations after the English clergyman-scholar Thomas Robert Malthus came to the dismal conclusions expressed in his 1798 "Essay on the Principle of Population," millions of Britons emigrated to the New World. Nevertheless, during that time the resident population of Great Britain doubled, doubled again, and almost doubled a third time!

One contention in Malthus's essay became widely known: "Population, when unchecked, increases in a geometrical ratio. Subsistence increases only in an arithmetical ratio."[1] Throughout the essay Malthus was referring to *human* population, and by subsistence he meant food. As we shall see, these conceptions were unduly narrow. But the really basic Malthusian principle is so important that it needs to be restated in the more accurate vocabulary of modern ecology. It states a relationship of inequality between two variables:

> The cumulative biotic potential of the human species exceeds the carrying capacity of its habitat.

No interpretation of recent history can be valid unless it takes these two factors and this relation between them into account. No one can truly understand the intensification of competition unless he grasps this principle. People whose political, religious, and ideological perspectives cause them to ignore these two variables will not succeed, even with the best of intentions, in improving the human condition.[2]

The phrase "biotic potential" refers to the total number of offspring a parental pair would be theoretically capable of producing. The cumulative biotic potential of the human species refers to the total number of people that could result after a series of generations if each generation fully exercised its reproductive power. The carrying capacity of the habitat, of course, is simply the maximum number of living individuals the available resources can indefinitely support. It is limited not just by the finite supply of food, but also by any other substance or circumstance that is indispensable but finite in quantity. The least abundant necessity will be the limiting factor; it may or may not be food. For industrially developed countries it began to appear in the 1970s that the limiting factor might be oil, while in some places it was water. It could be some other necessity.

For most of these seven generations, people fondly imagined that somehow Malthus had been proven wrong.[3] With the vast territorial expanse of the New World, and with new agricultural technology to exploit it, plus new means of transportation and trade, and improved organization, we expanded food production more than Malthus would have dreamed possible. Nevertheless, the Malthusian principle still held. We were only able to *suppose* it was mistaken because our exceptional circumstances (in the Age of Exuberance) made us commit two oversights. First, because the existing population was for the time being appreciably less than the world's suddenly augmented carrying capacity, we did not see the human relevance of the carrying capacity concept. Second, we failed to think about *cumulative* population. Until lately there always seemed to be room for the increment we expected next year, or in the next generation.

Charles Darwin was more perceptive. He recognized not only the Malthusian principle's truth, but also its significance. It was the key insight he needed to unlock the riddle of evolution—and thus to lay the foundation upon which his successors could build the science of ecology.[4] In effect, Darwin saw that the adjective "human" was unnecessary in the Malthusian principle, for the principle was not limited to one species. Darwin's version of it (recast here in the ecological language of today) was universal rather than specific:

> The cumulative biotic potential of *any* species exceeds the carrying capacity of its habitat.

As a result, there is *competition* among the members of a species population for use of resources that are in short supply relative to their numbers. Not all competitors will succeed; not every individual will live through all stages of the life cycle. The population, pressing on its limited resources, will suffer attrition.

The attrition of the population will not be random, however. Darwin saw that, since there are differences between individuals, some competitors will have advantages (however slight) that others will lack. Those with advantages will be somewhat more likely to survive to reproductive age; they will be somewhat more successful in mating, reproducing, and providing care for their progeny. Thus they will, on

127

the average, leave more descendants than those with disadvantages. Moreover, Darwin saw that what traits are advantageous and what traits are disadvantageous will depend on what environmental circumstances the organism must cope with. Environmental selection pressures, he realized, need not alter the traits of any individual. Evolution happens when environmental pressures merely influence comparative reproductive success. By influencing the comparative abundance of a particular organism's descendants, the requirements imposed by the habitat influence the future prevalence of that organism's inheritable traits among the total population.

The fact that evolution does operate shows that the Malthusian principle is valid. So, too, does the fact that there are food chains. If each species did not overreproduce, then predation would always lead to extinction of prey, the predatory species in turn would starve, and life would long since have vanished from the earth. Herbivores do not meticulously wait for seeds to ripen and fall before consuming the plants that bear them. Carnivores do not generously abstain from eating herbivores until the latter have replaced themselves with progeny. Every species that serves as another's sustenance can endure only by virtue of a reproductive capacity sufficient to compensate for such attrition. The persistence of life in a world where organisms consume other organisms confirms the Malthusian principle. To deny that principle, as we have naively done, is to deny evidence all around us.

With self-restraint, humans have been able sometimes to harvest such things as timber, fish, or other useful species on a "sustained yield" basis.[5] This fact should be seen as evidence for, not against, the Malthusian principle. Depletion of such resources by an overly prolific human species was never the only predictable result of the Malthusian principle. We supposed it was because we thought about the principle too anthropocentrically. Sustained yields represent reproduction in excess of replacement *by the resource species*; the excess is then "harvested" by an exploiting species—in this case, *Homo sapiens*. If Malthus were so wrong, there would have been no sustained yields of anything. In every bite of our daily bread there is a reminder of the wheat plant's ability to produce more seeds than required for its own replacement.

One of the great ironies of history has been the notion that *our* species was somehow exempt from a principle that manifestly applies to all other species. Malthus stated the principle of reproduction in excess of carrying capacity for man in particular. Darwin later generalized it to cover all species, and went on to discern its evolutionary implications. In the years since Darwin, most non-biologists seem to have smugly *reversed* Malthus by "slightly amending" Darwin's generalized version—accepting its application to all species except one, ourselves, the very species about which the principle was first asserted.

To be sure, man does differ from other animals, for man transmits to his descendants a cultural as well as a genetic heritage. The cultural heritage is conveyed socially rather than biologically—by symbolic language rather than by chemical code. It is thus more readily and deliberately mutable.[6] Its mutations can sometimes be mothered by necessity rather than by random accident, and perhaps somewhat more of them can be adaptive rather than lethal. But for some time there have been more and more signs that this cultural heritage does not exempt our species from the full ramifications of population pressure in a finite habitat. The ratio of load to carrying capacity can change culture itself. We had supposed our difference from other species exempted us. In the 1970s, some of the wistful remembrance of earlier times reflected the wish that we could go on supposing.

Learning How to See

People everywhere, but especially Americans and Europeans, had the illusion that the Age of Exuberance could be perpetual.[7] Our lives as individuals and as nations have been conducted in ways that would make sense only if living in a scientific age and knowing about evolution, food chains, etc., merely required us to believe that the cumulative biotic potential of *every species except Homo sapiens* exceeds the carrying capacity of its habitat.

In the 1970s we began—some of us—to open our eyes and look afresh at what was really happening.[8] To do this, we had to learn another lesson from Darwin, a lesson in how to see. Darwin was a meticulous observer of nature. He painstakingly collected vast arrays of information based on careful observation of natural specimens. But, like Thomas Kuhn more than a century later, Darwin also recognized that *one sees with ideas as well as with eyes.* He was fully aware that it was because he had absorbed an important idea (from Malthus) that he was able to see the pattern and meaning in his observations of nature. For Darwin, reading Malthus induced a paradigm shift.

The ideas by which people in a post-exuberant age need to see their world and their experience in it are the ideas of ecology. This means reinterpreting many things we have conventionally viewed in quite a different perspective. It does not mean we should merely become eco-pharisees—piously deploring the condition of Lake Erie, for example, and uttering newly fashionable words like "recycled" and "biodegradable" with little real comprehension, meanwhile ignoring the really basic process of succession.

Until recently, too few of us had learned nature's language, so we kept misreading the message implicit in our own experience. Our conventional perspectives have called attention to some facets of history that will be ignored in the pages that follow, but have caused us to ignore some facets we must now notice. The interpretation offered below, then, makes no pretense of being an exhaustive account of modern history. Nor does it purport to be a complete explanation for the events mentioned. It should be read merely as an attempt to review some familiar history in a way that fills some gaps in our usual understanding of how these things happened. This attempt emphasizes some ecological implications to which we have, in the past, been insensitive.

How to Read the Signs

The American way of life was generated by the tremendous contrast between a relatively unsettled Western hemisphere and the densely settled Eastern hemisphere.[9] Early in the nineteenth century, Americans committed themselves to the belief that these two hemispheres were two separate worlds, each with its separate destiny. The teaching of history in American schools held up the Monroe Doctrine as a vivid example of American wisdom; it had permanently declared the New World off limits to the Old World's rulers. Separate development of the two hemispheres became an article of American faith. Americans came to feel more and more detached from the "quarrels" of the Old World, while congratulating themselves for being blessed with unlimited opportunities for unquarrelsome progress in their vast undeveloped territory.

This faith in the separateness of the two worlds was transcended and yet also, in a sense, reaffirmed by the continuous flow of European immigrants to American shores. Each immigrant's decision to leave a benighted "old country" and seek his fortune in the land of the free seemed obvious testimony to the shining destiny of America. It provided what sociologists call "consensual validation" of the American dream. But an idea or a dream can be "validated" by consensus and yet be flagrantly invalid in terms of the conditions imposed by the real world. Americans were putting too much faith in the assumed permanence of a condition that had to be temporary. As people filled up the land, exploited its resources, and built great cities, the contrast between the New World and the Old was steadily diminishing. Americans were thereby undermining the American way of life by eliminating its environmental basis, just as countless other dominant species in other biotic communities have modified (adversely to themselves) the habitats upon which they depended.

Instead of recognizing these changes as symptoms of succession, public officials and dissidents alike persistently interpreted in political and moral terms events that needed to be viewed ecologically. Virtually everyone overlooked ecological causes of unwelcome changes. Our Age of Exuberance did not just terminate in the 1970s, or even in the 1960s. The exuberant phase was already shading into an excessive load phase as long ago as the time when American doughboys sailed for Europe to "make the world safe for democracy." That involvement with the Old World was a major departure from the separate development scheme for the two hemispheres. It implied that Americans had already begun to think of international relations as antagonistic (in the ecological sense of that word): the American people were coming to believe that their own opportunities to pursue happiness would be restricted by other nations' pursuit of other people's goals. I do not mean to imply that the antagonisms that were taking hold of us were imaginary. It is just that we were too deeply accustomed to interpreting them in political terms. We had not learned to understand the real nature and the real effects of the Age of Exuberance. Most people had never heard of such concepts as "carrying capacity" and did not think in terms of load limits.

At the close of the nineteenth century, the American culture of exuberance was deprived of the frontier conditions that had nurtured it. Distorted by this change, it expressed itself militarily in the very way Sumner had warned against in "Earth Hunger" (see Chapter 5). In the Spanish-American War, the United States, without quite knowing what it was doing, became an Old World type of political power by reaching out to take possession of some overseas territories. Previous American territorial expansion had involved contiguous lands that were expected eventually to become integral parts of the nation. This acquisition of more remote colonial possessions which were *not* intended to be nurtured to eventual statehood and full membership in the federal union marked an American shift to Old World perspectives.

The typical Old World method of reaping the advantages of the Age of Exuberance in spite of carrying capacity saturation at home had consisted of laying claim to overseas ghost acreage. The overseas colonies acquired by the United States never could function very effectively as ghost acreage, however; they were not rich sources of material wealth, as New World lands had been for the nations of Europe. So American imperialism was essentially fruitless (and it cost the United States a considerable loss of previously uncontested self-righteousness). Having acquired some remote colonial territory, the United States continued to profess but not really to practice separation of the two worlds.

It is difficult to believe that practical-minded Americans would have slipped so readily (and clumsily) into such self-deception if the growing settlement of their own continental territory had not already narrowed the difference between New World life and Old World life. Why else would the glorious little war fought against Spain for the ostensible purpose of consolidating separation of the hemispheres (liberating Cuba) have ended by the U.S. taking over the Philippines (hardly a part of the New World)?

Less than twenty years later, the German navy's "unrestricted submarine warfare" converted commerce between the New World and the Old into a force that involved America in what her citizens would formerly have termed "an Old World quarrel." Afterward, in their postwar recriminations, Americans failed to recognize that their commerce with Europe had *signified* (rather than caused) American abandonment of the concept that the two hemispheres were two worlds. Technological advances had increased the feasibility of trade across the oceans. Increased trade across the Atlantic showed that the two hemispheres were *not* two separate worlds. By praising such signs of progress as scheduled steamship traffic between Europe and America, and the transatlantic cable, for having "shrunk the globe," Americans were both rationalizing the termination of their professed commitment to hemispheric separateness and obscuring in their own minds the fact that they could never revert to the fortunate conditions of the Age of Exuberance.[10]

After the Great War the American people did try to revert briefly to the old notion of hemispheric separation. The United States stayed out of the League of Nations and piously proclaimed a firm intention never again to become involved in "pulling Britain's chestnuts out of the fire." Congress enacted "neutrality legislation" to keep the country from being drawn into another European war, but soon had to amend, evade, and then repeal it, when the American people found themselves again responding to the reluctant recognition that both shores of the Atlantic (and even both shores of the wider Pacific) were on the same planet. We tried going "back to normalcy" in other ways, too, but it turned out that disillusioned cynicism was simply unable to restore American faith in a New World with an exuberant and fully autonomous destiny. It could not have turned out otherwise, for the world had in fact changed; the ecological basis for the Age of Exuberance no longer existed.

Meanwhile, the continuing flow of immigrants into the land of opportunity had begun to be felt by earlier arrivals not simply as testimony to the American dream's validity, but now also as an economic threat. Pressure toward legal restriction of immigration was resisted, however, by political idealism that extended right up to the level of a presidential veto. Even a mind of the caliber of Woodrow Wilson's continued to see such issues in terms of idealistic principles unconstrained by ecological insights. He insisted on reaffirming the now anachronistic concept of America as a land of unlimited opportunity. America's former sparseness of population had enabled it so eminently to serve as a haven for the oppressed. Wilson (whose paradigm was pre-ecological) found it unthinkable that this might no longer be possible.

In the early part of the twentieth century, the number of arrivals from the Old World sometimes exceeded a million a year, and their competition in the labor market tended to keep wages from rising. Union leaders found the competitive presence of job-seeking immigrants an obstacle to organizing workers for collective bargaining. Eventually, in the 1920s, when labor union fears of competition for jobs coalesced with concerns of some misguided intellectuals about "mongrelization" of the American population by the increased immigration from southern and eastern Europe, a series of increasingly severe immigration restrictions became law. The ecological significance of the legislation lies in the fact that it did come in response to renewed intra-species competition. In the New World, where competition had once seemed to be mainly between man and other species, human niches had become saturated. From an ecological perspective, enactment of these restrictive laws can be seen as signifying the twilight of the New World's Age of Exuberance. But high-minded Americans continued for half a century to suppose it was more important to regard them as unequivocally shameful and to heap condemnation upon them, rather than to understand that such laws were expectable products of ecological change.

The racist element in the laws was real. Equally real were the increasing niche saturation and intensified competition, the facts that *aroused* exclusionist sentiments. In 1975, these sentiments were reasserted: 54 percent of Gallup poll respondents objected to admitting refugees from war-ravaged Vietnam *in a time of American unemployment*. President Ford, like Wilson, deplored such sentiments as an unworthy departure from America's "haven" heritage. A few years earlier, exclusionist sentiments had also appeared in Britain, in violation of a long tradition of Commonwealth-wide common citizenship. That tradition became *a casualty of British overpopulation,* and of the fact that ghost-acreage-supported Britain seemed a haven of freedom for a while to non-white immigrants from severely overpopulated former British colonies in Asia, Africa, and the West Indies. Britain's new immigration laws sought to soften the stigma of racism by also abridging the long-assumed "rights" of Canadians, New Zealanders, and Australians to settle in the mother country. The old idea that emigrants from Britain (and their descendants) could remain British, with full rights of reentry, simply was not viable in a post-exuberant world. Overseas Britons had been too exuberant and become too numerous, like everyone else.

In reporting these changes, news media continued to overlook the importance of the fact that Britain's carrying capacity was simply insufficient to accommodate everyone of British descent. The media indulged an illusion by exaggerating the causal role of politicians, such as the Conservative member of Parliament, Enoch Powell, who could sound so racist. Subsequently, even Australia and New Zealand, young nations though they were, felt obliged by economic difficulties to curb British immigration to their lands.

All these changes could have been seen as indicative of succession had the word been familiar and correctly understood. The exuberant expansion of European man into new niches overseas had wiped out the basis for exuberant expansion. The culture of exuberance stood in the way of recognizing this as an instance of succession.

Of course, it *is* deeply regrettable that Americans came to the point of closing their doors and excluding other would-be Americans, but it is also regrettable that they continued to misperceive the reasons for this change. Great ideals from the Age of Exuberance could hardly be reinstated in the populous world of the 1970s by bravely reopening the doors. Accordingly, ecological realists could hardly be anything but charitable in thinking about Britain when her dream of being "home" for all citizens of the Commonwealth was crumbling from erosion by the tide of almost three population doublings and heavy industrial growth since the time of Malthus. No habitat's carrying capacity is unlimited—not Britain's, or America's, or any other land's.

Environmental Brakes on Exuberance

Modern nations had staked their future on perpetual harvesting of supplies of non-renewable resources. Men now built with steel, concrete, or aluminum, rather than with wood. We had evolved into societies so large and complex that they required quantities of energy too vast to be supplied by contemporary crops of organic fuel. We allowed ourselves to become so numerous that we could not really grow the food we needed without enormous "energy subsidies," augmenting sunlight and muscle power in agriculture with industrially produced chemical fertilizers and fuel-burning machinery for planting, cultivating, harvesting, shipping and processing.[11] Americans thus were farming not only the great plains of Iowa and Nebraska but also the gas wells of Texas and Alaska and the oil wells of the continental shelf offshore. Even agriculture, the ultimate achievement in man's development of the takeover method of carrying capacity expansion, had become converted to drawdown methods. The most "prosperous" nations were living on phantom carrying capacity but had not learned the concept. By using still more enormous quantities of energy for new occupations unrelated to agriculture, *we put off recognizing* that our population had outgrown its maintainable niches. Had people understood the ecological implications of the Industrial Revolution, it might have been seen not so much as a great step forward for mankind but, as we shall make clear in Chapter 10, as a tragic transition to human dependence on *temporarily* available resources.

To provide charcoal for iron smelting, the British in the eighteenth century had harvested their timber faster than it grew. Depletion of British forests led to increased coal mining.[12] The need for pumping devices to remove water seeping into the mines provided the impetus for development of steam engines. Those engines could convert the chemical energy in a fuel such as coal into mechanical energy that could do useful work. In addition to working the pumps at the mines, other applications for steam engines were soon found, and they made great industrial expansion possible. Reliance on current photosynthesis (annual timber growth) was replaced by dependence on accumulated past photosynthesis (coal deposits). There was a rapid proliferation of coal-consuming machinery. As a result, Britain evolved an economy that traded British factory output for food and other agricultural products grown overseas. Those doublings of British population since Malthus, and the exporting of British people to other lands around the world, were thus made possible by exchanging a way of life based on visible acreage for a way of life based on *two levels* of ghost acreage. By heavy use of *fossil* acreage, British industry produced the goods that gave Britain access to *trade* acreage.

So British exuberance since Malthus was no refutation of the Malthusian principle. By the drawdown method Britons merely postponed the day of reckoning. They lived beyond their energy income, harvesting timber on a faster-than-sustained-yield basis. Then they bought further time by exploiting ghost acreage—both overseas and underground. Despite such measures, only two hundred years after James Watt's invention, the time Britain had thus bought seemed to be running out.[13]

Britain was the first nation to experience what economists came to call the "takeoff" into supposedly self-sustaining economic growth. Long after the evidence had turned against the idea, the illusion persisted that a similarly brilliant destiny remained possible for all nations. This was partly due to economic theorists' belief that the takeoff was a product of British accumulation of *monetary* savings in the form of profits from overseas trade.[14] It depended, in fact, on the geological accumulation of *energy* savings in the form of coal deposits.[15] In the last third of the twentieth century, although the vast majority of the world's energy savings were still in the ground, the vast majority of the most accessible fraction of those savings had already been extracted and spent.

In the decades since World War II, all the leading industrial nations had plunged still more deeply into living on nature's exhaustible legacy. Furthermore, they had committed themselves to technology that relied on the huge advantage petroleum had because it is liquid, whereas coal is solid. Consumption of petroleum products increased enormously, although petroleum was very much less abundant in the earth than coal. The relative share of energy obtained from coal diminished, and for some countries even the absolute amount declined. Many national economies made themselves deeply dependent on the continued sailing of growing fleets of supertankers. In the meantime, however, the volume of oil reserves discovered by each additional million feet of exploratory drilling was rapidly declining, despite expanded technical knowhow. This showed us that we had already extracted and burned the most accessible supplies, and that the existing and still increasing rate of consumption would virtually exhaust even the less accessible reserves within the lifetime of people already living. The social repercussions were going to be staggering.

Many people refused to believe it, and national economic policies continued to be based on a myth of inexhaustibility. The coal miners' strike in Britain in 1972 provided a temporary foretaste of the social and economic chaos that would have to come when a fuel-dependent world runs short of fuel.[16] The Arab oil embargo the following year added a sharp reminder. Before retiring as prime minister, however, the characteristically indomitable Harold Wilson looked to future North Sea oil "production" as Britain's economic savior—without acknowledging that salvation achieved by drawing down fossil acreage would have to be temporary.

Other mineral resources required by industry were also running shorter than people realized, were already probably operating as limits on the world's rate of industrialization, and may even have been already limiting the productivity of existing industrial facilities. In the United States, for example, the number of pounds of copper obtained *per ton of ore mined* was by 1965 less than half what it was in 1925. Over those four decades, total copper production had almost doubled; this had obscured for an uncritical public the problems connected with a fourfold increase in ore extraction.[17] Intensified efforts to secure such increasingly scarce resources could be expected to have serious social repercussions. Two cases will suffice to show the nature of the problem.

First, in Britain there were mounting worries over the likelihood that the remaining "amenity areas" of the country, such as Snowdonia National Park in Wales, would soon be devastated by strip mining for metallic ores needed by British and European industry.[18] For many metals, the world's richest ores had already been mined and smelted. Leaner ores had begun to be in increased demand. The leaner the ore being mined, the greater the volume of rock removed per ton of metal obtained. Even if the budget for mining operations were required to include provision for cleaning up and replanting after an area had been stripped, it would be impossible to restore the landscape characteristics that were imparted to Britain by the Ice Age. These were the features that contributed so much to the beauty of Britain's northern countryside.

Second, in an effort to prevent white minority-ruled Rhodesia from establishing its independence in disregard of the full rights and aspirations of its large black majority, the United Nations had imposed economic sanctions upon that country. UN member nations were obliged to refrain from importing Rhodesian goods. But, faced with difficulty in obtaining from anywhere else (except Russia) adequate supplies of chromium needed for certain alloys, the United States government chose in 1972 to stop complying with the UN sanction. American firms resumed importing Rhodesian chrome.[19] The American government decided to permit this in spite of strong internal pressure against giving aid and comfort to a regime internationally branded as racist.

If, as these examples show, advanced industrial nations with humane democratic traditions were thus pressed toward actions deeply deplored by many of their own citizens, it was clear that national economies were already feeling the pinch of ecological limits. These limits were the ultimate basis of some of the most revolutionary changes in our lives, particularly the rising sense of post-exuberant despair and oppression. The number of resources upon which our way of life depended for its continuation had been very greatly enlarged by the ecologically distinctive traits of modern humans. So the depletion of almost any one of many resources that were now essential could put the brakes on our neo-exuberance.

The Real Error

Malthus did indeed err, but not in the way that has been commonly supposed. He rightly discerned "the power of population" to increase exponentially "if unchecked." He rightly noted that population growth ordinarily is *not* unchecked. He saw that it was worth inquiring into the means by which the exponential growth tendency is normally checked. He was perceptive in attaching the label "misery" to some of the ramifications of these means. Where he was wrong was in supposing that the means worked fully and immediately. (That *this* was his error has not been seen by those who reject his views.)

Being himself under the impression that it was not possible for the human load to exceed the earth's carrying capacity, Malthus enabled those who came after him to go on misconstruing continued impressive growth as evidence against, rather than as evidence for, his basic ideas. Carrying capacity was a concept almost clear to Malthus. He even sensed that the carrying capacities of earth's regions had been repeatedly enlarged by human cultural progress.[20] If he was not yet able to make clear to himself and his readers the distinction between means of enlarging carrying capacity and means of overshooting it, we do ourselves a serious disservice by perpetuating his shortcoming. And we do just such a disservice by continuing to mistake overshoot for progress, supposing drawdown to be no different from takeover. By erring thus we prolong and deepen our predicament.

Despite Malthus's belief to the contrary, it *is* possible to exceed an environment's carrying capacity—temporarily. Many species have done it. A species with as long an interval between generations as is characteristic of ours, and with cultural as well as biological appetites, can be expected to do it. Our largest per capita demands upon the world's resources only begin to be asserted years after we are born. Resource depletion sufficient to thwart our children's grown-up aspirations was not far enough advanced when our parents were begetting, gestating, and bearing us to deter them from thus adding to the human load.

By not quite seeing that carrying capacity *can* be temporarily overshot, Malthus understated life's perils. He thus enabled both the admirers and the detractors of his admonitory writings to neglect the effects of overshoot—environmental degradation and carrying capacity reduction. In his analyses he assumed linear increase of carrying capacity. While this fell short of sustaining exponential growth of would-be consumers, it was, even so, a far brighter prospect than carrying capacity *reduction*.

Notes

1. Malthus 1906, p. 7. Note the qualifying phrase, "when unchecked." Just as Isaac Newton had not completed the statement of his famous first axiom (which began, "Every body continues in its state of rest, or of uniform motion in a right line") until he added the qualifying clause, "unless it is compelled to change that state by forces impressed upon it," so Malthus meant the qualifying phrase, "when unchecked," to be taken as an essential ingredient of his axiom. Newton was not trying to prove that all motions are straight line motions; he was trying to establish a basis for analyzing how and why motions depart from straight lines. Likewise, Malthus was not trying to show that population growth, as observed, always follows an exponential curve; he was trying to establish a basis for analyzing why it doesn't, and for understanding the real social significance of the forces impressed upon it.

2. Allan Chase, for example, denounces as the "core myth" of Malthus the idea that our power to reproduce exceeds our power to provide sustenance. In attempting to develop his argument that Malthus was biased against the poor and against other races, Chase neglects passages in Malthus's essay that show he was not merely defending an Anglo-dominated status quo. See Chase, *The Legacy of Malthus: The Social Costs of the New Scientific Racism* (New York: Alfred A. Knopf, 1976), pp. 77–79. For example, Malthus (1906, p. 83) wrote: "There are many parts of the globe, indeed, hitherto uncultivated and almost unoccupied; but the right of exterminating or driving into a corner where they must starve, even

the inhabitants of these thinly-peopled regions, will be questioned in a moral view." He was neither morally myopic nor misanthropic when he perceptively pointed out (1906, p. 83) that "If America continue increasing, which she certainly will do, though not with the same rapidity as formerly, the Indians will be driven further and further back into the country, till the whole race is ultimately exterminated." Nor was he laying foundations for racist policies by also insisting, as he did, that "To exterminate the inhabitants of the greatest part of Asia and Africa, is a thought that could not be admitted for a moment." By personality, Malthus was, according to those who knew him, cheerful, serene, amiable, good natured, and gentle (see Appelman 1976, p. xi). His purpose in writing about the vice and misery by which population pressure would tend to be checked in the absence of moral restraint was to enhance the probability that moral restraint would obviate the otherwise inexorable drift toward even those kinds of vice and misery deplored by writers like Chase.

3. This has been especially the case, particularly in the twentieth century, among economists. See the discussion in Glass 1953, pp. 60ff,; and for an exemplification see Hagen 1962, pp. 47–48. Blaug (1968, p. 551) claims that at the turn of the century every schoolboy "could prove that Malthus had gone wrong by underestimating both the potentialities of technical progress and the possibility of family limitation by birth control devices," and he insists not only that this "refutation" is "perfectly valid" but that it can be surpassed in sophistication when Malthusian theory is recognized as "metaphysics masquerading as science."

4. See Appelman 1970, pp. 71, 83–84, 103, 117, 492.

5. As was pointed out in Ch. 3, note 9, carrying capacity for a predator (or consumer) species can be defined as the maximum population that can be permanently supported by sustained yields from prey (or producer) species.

6. See the chapter on "The Evolution of Human Behaviour" in Burnet 1971.

7. For example, see Hagen 1962, Chs. 2, 3, and Appendix II; Horowitz 1969. Cf. Culbertson 1971, pp. 29–32, 293.

8. See Culbertson 1971, preface and Chs. 1, 15.

9. See references to Sumner, Turner, and Webb, in Chs. 2, 3, 5.

10. For a more complete, but also more conventional (i.e., non-ecological) treatment of the events of American history recounted in this and following paragraphs, see Morison 1965 (listed among references for Ch. 5).

11. See, for example, Steinhart and Steinhart 1974, and Udall et al. 1974 (both listed among references for Ch. 11); also see Cook 1971.

12. Cottrell 1955 (listed among references for Ch. 3), p. 79.

13. Goldsmith 1971; cf. Taylor 1970.

14. See Culbertson 1971, pp. 241–243, 281–286.

15. Cottrell 1955 (see note 12, above), pp. 124ff.

16. See numerous articles in the London *Times* from Jan. 25, 1972, through February, and on into March.

17. Lovering 1969, p. 124.
18. James Lewis, "Mining and the National Parks," *The Guardian,* Apr. 1, 1972.
19. See editorials in the *New York Times,* Jan. 26, 1972, p. 36; Feb. 28, p. 30; Mar. 17, p. 40.
20. This is evident in the following passages from the 1798 Essay (Appelman 1976, pp. 27, 29, 31): "In the rudest state of mankind, in which hunting is the principal occupation and the only mode of acquiring food, the means of subsistence being scattered over a large extent of territory, the comparative population must necessarily be thin." "It is well known that a country in pasture cannot support so many inhabitants as a country in tillage. . . ." "The reason that the greater part of Europe is more populous now than it was in former times is that the industry of the inhabitants has made these countries produce a greater quantity of human subsistence."

Selected References

Appelman, Philip, ed.
1970. *Darwin: A Norton Critical Edition.* New York: W. W. Norton.
1976. *An Essay on the Principle of Population: A Norton Critical Edition.* New York: W. W. Norton.
Blaug, Mark
1968. "Malthus, Thomas Robert." Pp. 549–552 in David L. Sills, ed., *International Encyclopedia of the Social Sciences,* IX. New York: Macmillan and Free Press.
Burnet, Macfarlane
1971. *Dominant Mammal: The Biology of Human Destiny.* Harmondsworth, Middlesex: Penguin Books.
Cook, Earl
1971. "The Flow of Energy in an Industrial Society." *Scientific American* 224 (Sept.): 135–144.
Culbertson, John M.
1971. *Economic Development: An Ecological Approach.* New York: Alfred A. Knopf.
Darwin, Charles, and Alfred Russel Wallace
1958. *Evolution by Natural Selection.* Cambridge: Cambridge University Press.
Glass, David, ed.
1953. *Introduction to Malthus.* London: Watts.
Goldsmith, Edward, ed.
1971. *Can Britain Survive?* London: Tom Stacey.
Hagen, Everett E.
1962. *On the Theory of Social Change: How Economic Growth Begins.* Homewood, Ill.: Dorsey Press.

Horowitz, David
  1969. *The Abolition of Poverty.* New York: Frederick A. Praeger.
Lovering, Thomas S.
  1969. "Mineral Resources from the Land." Ch. 6 in Committee on Resources and Man, *Resources and Man.* San Francisco: W. H. Freeman.
Malthus, Thomas Robert
  1906. *Parallel Chapters from the First and Second Editions of An Essay on the Principle of Population, 1798, 1803.* New York: Macmillan.
Taylor, L. R., ed.
  1970. *The Optimum Population for Britain.* London: Academic Press.

9 | Nature and the Nature of Man

Detachable Organs

Somewhere in North Dakota I had an unexpected opportunity to discover just what is most ecologically distinctive about *Homo sapiens*. My family and I were on an automobile trip, driving east on a two-lane road. We had had to slow down appreciably when we became momentarily trapped behind a bicycle. Soft shoulders at the side of the road confined the cyclist to the pavement, and westbound traffic in the other lane prevented us from passing him for a mile or so.

While driving our car for a few moments at the unfamiliar speed of the bicycle ahead of it, I began to notice details of this "traffic obstruction" that otherwise would have escaped my attention. First, because sporadic fuel shortages had already raised the possibility that automobile touring might someday be only a fond memory, I studied the pannier bag hanging astride the rear wheel of the bike, and wondered just what items of spare clothing, etc., one might be able to include in one's severely limited luggage when traveling on two wheels instead of four. I noticed also the muscles in the cyclist's tanned calves, and the steady rhythm of his pedaling. Then a slight motion of his head called my attention to a short, stiff wire that curved forward from the bill of his plaid cap. On the tip of the wire was a small round object which I suddenly realized was a thumbnail-sized mirror. Being just a few inches from his face, a mirror small enough to offer negligible wind drag was large enough to give his inboard eye an appreciable field of vision to the rear, without requiring him to turn his neck awkwardly and unbalance himself in looking back over his shoulder.

To all intents and purposes, I realized, this cyclist had "eyes in the back of his head." Before an opportunity arose to speed on past him, another realization had become obvious—that having easy and adequate rear vision could have significant survival value for anyone pedaling a bike on a highway used mostly by automobiles. Then the phrase "survival value" brought Darwin to mind.

Even at automobile speed, crossing the North Dakota prairie gives one time for extended thought. Long after the bicycle tourist was no longer visible in my own rear-view mirror (an artifact I had taken for granted for years), I was contemplating the thousands of generations it might have taken for unfavorable mutations to be eliminated and the rare ones that yield advantageous phenotypic traits to accumulate through natural selection to the point where humanoid creatures might literally have had an eye or two facing aft. I knew it could not happen. Even if humans lacked the inventiveness to short-cut the process by hanging a mirror from a cap, the era in which bicycles would be followed by automobiles was not going to last through enough generations for selection pressures from this situation to influence evolution in that way.

But rabbits had done it! I remembered learning in my childhood that the eyes of a rabbit, positioned at the sides of its head, have fields of vision that overlap slightly in the rear, rather than in front—an evolutionary result of what happened to rabbits if they could not always see any predator pursuing them while they leaped evasively this way and that, heading for cover.[1] For the rabbit, rear vision given him by evolution entailed only minor costs. By slight movements of his head, the rabbit's effective field of sight could span the full 360 degrees, and the fact that his eyes did not face straight ahead was no disadvantage. The rabbit's mode of living did not require the precise depth perception achieved by simians and humans through placement of both eyes in a frontal position permitting stereoscopic vision.

For monkeys, however, there was less survival value in rear vision than in accurate depth perception. To move advantageously through an environment of tree branches, where misperception could mean a fatal fall, required not only a restructuring of mammalian paws (into grasping hands) but a restructuring of the head so that converging eyes would enable distances from branch to branch to be perceived with centimeter precision.[2] An arboreal niche had imposed selection pressures that caused the useful combination of hands and binocular vision to evolve to high perfection. Later, some descendants of creatures thus endowed somehow became surface dwellers. In a new niche, the advantages of depth perception and grasping hands could be turned to new uses. Eventually the result was a proliferation of tool-using. Tools (including weapons) enabled the descendants of tree-dwelling simians to do on the ground things they could not have done so well with only the organs their genes gave them.[3] Chipped stones could cut things teeth couldn't sever—or could be replaced when broken teeth could not. Hollow gourds (and later mud-daubed woven baskets, and still later fired pottery) could carry larger quantities of water a greater distance from the riverbank than a pair of cupped hands could.

Thus there came to be a population of creatures whose way of living involved considerable use of external supplements to their own organic equipment. Insofar as they found it advantageous to use such detachable organs (for that is what tools of any kind are), these creatures became subject to new selection pressures. Now that tool-using was part of their method of coping with their habitat, those with greater ability to devise and use tools would be able (on the average) to leave more progeny than those with less such ability. In short, there was now going to be selection based on ability to make and use detachable organs. Before there had always been selection only for organs (or behavior patterns) capable of coping with the habitat but integral to the creature's body.

Man, the Prosthetic Animal

Nature had achieved, so to speak, an evolutionary breakthrough. Human extensions of natural organs have come a long way since teeth began to be supplemented with flint scrapers. Some of the supplementary artificial organs have even moved inside. Anyone with a denture (or even a filling) in his mouth can appreciate the technical advances that have, in effect, made teeth repairable or replaceable. Defects or damage in other organs can also be redeemed by artificial means. Anyone who wears glasses can easily recognize the utility of detachable organs, though he may never have thought to apply such a phrase to the device he hangs on the bridge of his nose and hooks behind his ears each morning. His visual acuity depends upon external supplements to the imperfect organic lenses within his eyes.

For a small minority of the human species, the availability of more sophisticated detachable organs has been more than a matter of mere convenience. It has been a matter of life or death. Several good friends come to mind whose hearts have been stopped and surgically repaired. Not only were the tools wielded by the surgeon (as extensions of his hands) vital to these patients; their lives were sustained during the operation by heart-lung machines—temporary substitutes for their own hearts and lungs. Thousands of people have also been enabled to live on after complete failure of their natural kidneys by attaching themselves periodically to a machine that externally performs the blood-purifying hemodialysis no longer possible in their own bodies.

In a milder way, an artificial leg is a tool designed to perform at least some functions of a natural limb. The word "prosthetics" refers to the branch of surgery that deals with replacement of missing parts.[4] It may seem a rather remote and morbid subject, but all humans inhabiting other than tropical environments are users of essentially prosthetic devices—clothing. *Homo sapiens* can survive where he would otherwise perish by taking from another animal the protective cover normal to that animal as a product of natural selection. Likewise, our shoes have served as a kind of prosthesis for the hooves we were not equipped with at birth, enabling us to walk additional portions of the uneven face of the earth. The evolutionary and ecological significance of such prosthetic devices has been to facilitate the spread of mankind over a more extensive range than we could have occupied with only the equipment of our own bodies.[5]

Eventually man learned to obtain prosthetic insulation without killing the animal source—by shearing wool, spinning it into yarn, and weaving or knitting it into garments. More recently it even became possible to do the spinning, weaving, and knitting without involving other (contemporary) animals at all, relying on the remains of organisms that lived during the Carboniferous era by using fibers chemically synthesized from coal or petroleum derivatives.

The ecological implications of prosthetics have not been clearly seen by either social scientists or biologists because the concept has seemed so narrowly medical. Yet nature's evolutionary breakthrough may best be understood by viewing it as a shift from (a) selective retention of *organic traits* on the basis of their adaptive utility to (b) selective retention of *prosthetic tools* on the basis of their adaptive utility. It was no longer necessary to evolve eyes in the back of the head when a species needed rear vision; a tool could cope with the problem.

Viewing tools as detachable organs, viewing man as the prosthetic species, and viewing man-tool complexes as a new component of ecosystems were perhaps novel ways of looking at familiar things, but they were not really radical departures from conventional evolutionary knowledge. Nature had made comparable evolutionary breakthroughs before; these were known already. The resemblance could be recognized and could be illuminating.

Life had once been confined to the sea. The chemical processes involved in life depend on moisture. Invasion of dry land by living forms could come about only as more or less waterproof skins were evolved that enabled organisms to maintain within themselves a wet "internal environment" even when not immersed in water. By enclosing a bit of the sea within themselves, organisms could escape from the sea.[6] Hundreds of millions of years later (but still scores of millions of years ago) a similar principle was applied again by nature in opening up another new range of niches. When animals had depended largely upon the temperature of their surroundings as the main determinant of the temperature inside their skins, opportunities for animal life remained scarce wherever environmental temperatures rose to tissue-damaging highs, or fell to activity-suppressing lows. The evolution of "warm-blooded" birds and mammals opened new niches in new climates by enclosing within an animal's skin a homeostatically temperate mini-climate.

Man has further extended this enclose-and-control principle of nature, not only with clothing, but also with centrally heated or air-conditioned buildings. If the mini-sea or the mini-climate enclosed within the skin of an organism is ordinarily regarded as part of the organism, rather than part of its habitat, then, broadly speaking, even furnaces, refrigeration units, and thermostats can all be understood as prosthetic devices. They are detachable parts of *extended* human beings. The walls of our buildings and the shells of our vehicles are (like the Eskimo's fur parka) a kind of prosthetic skin with which we surround our enlarged selves, enclosing mini-environments within them.

When nature spawned man, nature generated a new mode of interaction between organism and habitat.[7] Man would invent new ways of enclosing within his own expanding boundaries those conditions required by his original organic traits. Much of what would happen after nature made the human breakthrough would amount to spontaneous and prolific exploration of the potential of that new mode of interaction between organism and habitat. By the time the Age of Exuberance turned into a post-exuberant age of overpopulation, it was beginning to be possible to see not only the amazing lengths to which the prosthetic nature of man had carried nature's experiment, but also the fact that the enormous potential of the new mode was not without both hazards and limits.[8]

Species of Many Niches

Several years before I happened to see the mirror-wearing cyclist on the highway in North Dakota, I had attended in Boston a meeting of the American Society of Naturalists. The presidential address was given by a California geneticist, G. Ledyard Stebbins.[9] His topic was "The Natural History and Evolutionary Future of Mankind." He had looked at the human species in the way naturalists look at other species, raising a more or less standard set of questions about it. Where does man occur, he asked. Everywhere on earth, he answered. Why does man occupy so ubiquitous a niche, he asked. Because he can modify his external environment, he answered. When did he occupy this niche? From many thousand years ago to the present. How? By using tools, organizing himself into societies, and coping with problems by means of deliberative foresight. Man's future, said Stebbins, would depend on improving his capacity for social organization and overcoming his tendency for self-deception; he sometimes pursues attractive but disastrous goals.

It was a thought-provoking address, but I had gone away bothered by his reference to man's niche (singular). As a sociologist, I was trained to be sensitive to the diversity of man's ways, and it seemed to me to make better sense to think of our species as filling *many* niches.

Especially in modern industrial societies, human life is characterized by a highly ramified division of labor between a vast assortment of occupational specialties. This has had adaptive advantages. If every author had to grow his own wheat and bake his own bread to obtain the strength to mine his own ore and smelt his own iron to make his own typewriter before starting to write, the world would have few typewriters and few books (even on agriculture, baking, mining, or manufacturing). The cliché fits: if people had to be jacks of all trades, they would be masters of none—and there could be only a few trades and comparatively few people.

What happened was that the members of a single species discovered ways to behave almost as if they were many different species. Technology facilitated this useful differentiation of one species into many "quasi-species." A man or woman with one set of tools could do one sort of job (fill one sort of niche), while a person with a different set of tools could do another sort of job (fill another kind of niche).

Long after I heard Professor Stebbins's address, the mirror on the cap of the bicycle tourist gave me closure! It occurred to me that the choice between speaking of one worldwide niche versus speaking of many distinct niches scattered across all parts of the planet simply depended on whether tools were seen as devices that modify habitats to fit them to people, or as detachable modifications of people that fit them to varied habitats.

Highly varied man-tool combinations can live in highly varied environments; that was the fact to which Stebbins had called attention. It could be interpreted two ways, however. From the one-niche perspective, tools could be said to have enabled *Homo sapiens* to homogenize an otherwise diverse world—making all parts of it available for human habitation. From the multiple-niche perspective, the same tools could be said to have enabled a more or less homogeneous species to diversify itself enough to live in many different kinds of habitat. To a cultural anthropologist or a sociologist, the latter interpretation was more congenial. *Homo sapiens* remains one species genetically, but by means of diverse sets of artifacts (and diverse customs for using them) this one species has become ecologically diversified. Social scientists carried their insight too far, however, as we now begin to see. They so emphasized the power of mankind to become diversified culturally (including technologically) that they assumed this differentiated *Homo sapiens* from all other species and made irrelevant the concepts by which we understood the forces affecting other forms of life. Sociologists, impressed with the special humanness of humans, almost lost sight of the mutuality of impact of organism and habitat upon each other. Regaining awareness of the *interaction* of organism and habitat was a necessary antidote to our self-deceptive tendency, which, as Stebbins pointed out, has enabled our species to pursue some disastrous goals (such as exuberantly overshooting permanent carrying capacity).

All organisms have to adapt; it has always been arguable whether adaptation refers to what the habitat does to the organism, or what the organism does to the habitat. Always the effect is really mutual. As we saw in the last section of Chapter 7, recognition of the mutuality of the effect is the distinctively ecological insight that has been so difficult for the human mind, after an Age of Exuberance, to keep in focus.

Adaptation means something has been *modified* to "fit" or "go with" something else. I had seen the mirror on the cyclist's cap as a modification (extension) of his visual organs, not as a modification of his environment. In the same way we could see flint scrapers as supplementary teeth or fingernails. Proto-humans with flint scrapers could interact with their environment in ways that other proto-humans with only natural teeth and fingernails could not. After thousands of generations, these most primitive tools would be superseded by more sophisticated supplementary organs. The pocketknife I have carried since I was in grade school, for example, excelled the flint scraper's utility by as much as that stone tool had excelled nature's tooth and claw.

Not only would man devise artifacts that could do better the things his own organs might have done crudely; he would also devise them to do things he could not otherwise have done at all. Just as we did not have to evolve eyes in the backs of our heads when the need arose to see behind us, we did not need to evolve feathered forelimbs in order to fly. Most of us easily took up the habit of assuming that we fly by "getting into" an airplane. But when an airline pilot with thirty-three years of flying experience refers to the familiar act of buckling his cockpit seatbelt as "strapping a DC-8 to my waist," it is clear that even a modern jetliner can be seen as an elaborate prosthetic device. From the one-niche perspective, we might have thought of it as a tool for modifying the thin upper atmosphere to make it inhabitable by surface-dwelling man. Wings speeding through thin air make it "solid" enough to support the passengers' weight, and the cabin pressurizing pumps make it dense enough for them to breathe. From the multi-niche perspective, the airplane is a set of prosthetic wings and lungs that people "put on" when they have occasion to fly into an environment for which they were not genetically prepared.

Looking at matters this way, it was not just Neil Armstrong who in 1969 took the "short step for a man" down from the modified environment within the lunar landing craft onto the moon's inhospitable surface. It was a fantastic man-tool combination (astronaut clad in spacesuit with life support pack) that took the "giant step for mankind." By the enclose-and-control principle, this man-tool combination was recapitulating the experience of the sea-bred creature that first made a go of it on land.

Altered Carrying Capacity

Two million years after his ancestors learned to walk upright, *Homo sapiens* (with tools par excellence) had left footprints on the moon. He had also learned to fly in bunches of three or four hundred to business conferences on the far side of a continent, or to holidays on the far side of an ocean, by strapping on a DC-10 or a Boeing 747.

If people in the post-exuberant world could really become accustomed to thinking of even a jumbo jet as an enormous prosthetic device, a detachable organ, perhaps they could then make headway toward understanding the real ecological nature of the predicament our species had got itself into.

As we have seen, *Homo sapiens* made a breakthrough ten thousand years ago that enlarged (for the time being) the carrying capacity of his habitat. He learned to plant seeds and manage the growth of vegetable matter for his own sustenance, rather than leaving it to untended nature. As his horticultural endeavors progressed, advancement had to include improved methods of cultivating the soil. The simple digging stick was superseded by the hoe and eventually by the ox-drawn wooden plow. Only one hundred centuries after agriculture began, the digging process was accomplished in the most "advanced" nations by steel plows propelled by tractors that burned fossil fuel.

In those same societies, *Homo sapiens* had in the meantime discovered other reasons for digging into the earth's surface. Besides wanting to plant seeds, people now wanted to extract from below the surface vast quantities of non-renewable resources such as coal. If the digging stick was a prosthetic device, so was the modern power shovel, however enormous it might have become. Modern man's dependence on the ghost acreage of fossil fuels was so great that his power shovels became incredibly huge. A Kentucky lawyer, writing in *The Atlantic Monthly* of September, 1973, described industrial man as "the greatest geological calamity to hit the world since the melting of the ice sheets." He portrayed one of the mining devices operated by the Central Ohio Coal Company. The machine stood twenty stories high, he said. It used five-inch steel cable to lift a boom 310 feet long. With each bite into the earth it could remove 325 tons of "overburden" to permit "recovery" of coal underneath. [10]

From the time when the evolutionary breakthrough by nature brought forth the human way of evolving, the carrying capacity of man's habitat has never been fixed. Ever since man became human by making himself dependent on tool use, his dependence has consisted increasingly of the fact that he occupied niches made possible by his tools—niches that were previously nonexistent. In broadest terms, then, the function of technology has been to enlarge the human carrying capacity of any habitat. From dugout canoe to jet clipper, and from digging stick to giant power shovel, technology had enabled human beings to go places they could not otherwise reach, and to use substances they could not otherwise have exploited. Without the technology to make them usable, many substances in man's environment would not have become "resources."

Too Much of a Good Thing

Technological progress went beyond facilitating the takeover method of carrying capacity expansion. It led to our commitment to the self-destructive drawdown method. It facilitated useful quasi-speciation, but it also fostered *excessive* quasi-speciation, as we shall see in a moment. Nothing inherent in tools guaranteed that they would always be used in positively adaptive ways. The capacity for self-deception mentioned by Stebbins was important. Tools could be misused, and tool-using could also lead to an overwhelming "success" that would be tantamount to failure.

Consider misuse first. Take as an example the fire-making and fire-using tools that were from the outset an important part of human technology. These tools helped mankind multiply and spread into new niches. They extended the list of organic substances people could use as food. They extended the list of inorganic substances they could use for making other tools. They supplemented the heat-generating ability of a warm-blooded, mammalian body.[11] Eventually they became sophisticated enough to give access to that temporary carrying capacity buried as ghost acreage in the Carboniferous period. But always there was the possibility of accidental or deliberate misuse of fire. Fire could injure. Fire could destroy tools; it could destroy crops; it could destroy forests people thought they were not using (because they had not learned to recognize the indirect ways in which such "unused" lands served them). After a human-caused forest fire, with water-retention capacity lost in the denuded area, erosion could result that would destroy the crop-growing potential of lands people *did* know they were using.

But consider also the fact that essentially symbiotic (and even predatory and parasitic) relations were generated among human beings when their tool-using cultures enabled them to differentiate into many interacting "quasi-species." Another form of misuse arose and became common as a result of this. Each person's welfare no longer depended simply on what he could himself harvest from soil and water. It now depended on what he could obtain by exchange, from other human beings differently specialized. Whatever product or service one's own specialized efforts produced, it was unlikely one could subsist on it alone. A person had to obtain from others fulfillment of some of his own needs, and provide fulfillment for some of theirs. That was a basic aspect of being human. Because of the exchange relationship generated by this aspect of humanness, human beings came to be defined as resources. When humans defined humans as resources, misuse became probable.

In nature it was normal for one species to be a resource to another. Lions ate zebras not because lions were vicious, but because lions were lions (and zebras were edible). Man, the prosthetic species, was capable of (or subject to) quasi-speciation as different men used different tools. As a result, men would exploit (use) other men simply because men were men (and other men were different).

Homo sapiens was a species that could enlarge the carrying capacity of its own habitat by using tools. This ability meant that human beings would increasingly discover ways of using *as* tools things that had not been tools originally—including other human beings. Just as a non-lactating mother could use a cow to provide milk for her child, or a blind person could use a dog to lead him about, so one human being could use another human being's skills.[12]

One organism can be a prosthetic device to another. Being socially differentiated, a human population could be prosthetic to one another, although humans were never *merely* prosthetic devices. Cows and dogs generally acquiesced in their tool roles, but humans might not. If it was natural for humans to function as tools, it was also natural for human tools to assert non-tool aspects of their humanity. Inherent in man's capacity for quasi-speciation, therefore, was a source of chronic tensions. Prosthetic differentiation of the human species was a cause of human woe far more deeply endemic than either the "contradictions" Marxists supposed were specific to capitalism, or the aggressive and deceitful tendencies "free world" leaders supposed were specific to authoritarian regimes.

The very aspect of human nature that enabled *Homo sapiens* to become the dominant species in all of nature was also what made human dominance precarious at best, and perhaps inexorably self-defeating. Although man could not be human without quasi-speciation, perhaps with it he could neither avoid habitat destruction nor remain *humane*.

Which brings us back to the problem that success can lead to failure. Mankind extended to an unprecedented scope nature's principle that niches can be multiplied by the technique of enclose-and-control. Human beings became enormously numerous by using a vast assortment of detachable organs, many of them fantastically complicated and prodigiously large. An almost incredibly large fraction of the biosphere was thus enclosed and controlled by man-tool systems. This situation had to bring our species up against the influence of limiting factors. Belief in the illusion of limitlessness could not protect us; in fact, it probably hastened our encounter with nature's limits by hastening drawdown.

For all organisms, the necessity of maintaining some constancy of conditions inside the system has required the existence of an "outside." Maintenance of constancy within the system involves processes of importing and exporting materials and energy—to compensate for changes that would otherwise happen inside the organism's boundaries. Life has always depended on opportunities for living systems to obtain "resources" from their environment, and to exhaust spent substances (with autotoxic properties) into the environment. So the advantage to be gained by enclosing a piece of the environment within the system has always depended on there being plenty of environment left outside. Too many systems too much enlarged vitiate this advantage by seriously diminishing the environment/system ratio. (To those who use the word "exploitation" politically rather than ecologically, the importance of all this remains invisible.)

Homo sapiens has exploited too much. Human "success" entailed enclosure of an unprecedentedly large fraction of the total environment within the expanding boundaries of proliferating man-tool systems. There was consequently less "outside" in proportion to "inside" than had ever before been the case. The ironic result was that *technology, which originally had been a means of increasing the human carrying capacity per acre of space or per ton of substance, became instead a means of increasing the space required per human occupant and the substance required per human consumer.*

Post-exuberant man found himself playing a game with changed rules. It was no longer uniformly true that additions to technology added to his habitat's carrying capacity. At 500 miles per hour it took fewer planes than at 150 miles per hour to make the airways crowded. At 325 tons per bite, it took fewer digging operations than with a digging stick to devastate the countryside.

Man used to live in a world where carrying capacity was equal to the product: resources *times* technology. Man's "success" changed it into a world in which carrying capacity was coming to equal the quotient: resources *divided by* technology. The predicament of mankind no longer consisted merely of the simple Malthusian problem of an expanding population pressing against *fixed* limits of a finite habitat (or against less rapidly rising limits of a stretchable habitat). Now it was a worse predicament—an expanding population with burgeoning technological power was *shrinking* the carrying capacity of its habitat.

The human community was undergoing *succession* as a direct

consequence of man's nature as a prosthetic or tool-using species. The more potent human technology became, the more man turned into a colossus. Each human colossus required more resources and more space than each pre-colossal human. Contrast the environmental impact of the Central Ohio Coal Company and its huge machines with the environmental impact of the Stone Age people who inhabited the same area a few centuries before.[13] The Indians had not necessarily possessed any more virtue; they simply used cruder tools. They were non-colossal.

The same kind of problem would be much easier to recognize if mankind were afflicted with some kind of mutation that had the curious effect of causing children to grow to twice their parents' adult size—so that they required twice as much food and fiber per capita to sustain life and comfort. Suppose, further, that the effect were somehow cumulative, so that each generation grew twice as large and voracious as the preceding one. Quite obviously, the world's carrying capacity would be much less for later generations of giants than for earlier generations of runts. Just as obviously, neither conventional political nostrums nor revolutionary agitation could do much to remedy the situation.

Perhaps we would know this if we ceased to call ourselves *Homo sapiens* and began to call ourselves *Homo colossus*. If we were accustomed to thinking of a human being not just as a naked ape or a fallen angel but as a man-tool system, we would have recognized that progress could become a disease. The more colossal man's tool kit became, the larger man became, and the more destructive of his own future.

Notes

1. Actually, the visual fields of a rabbit's two eyes do overlap somewhat in front; the rear overlap is not in the horizontal plane, but is somewhat elevated, suggesting that selection pressure from predatory birds may have had more to do with rabbit evolution than threats from surface-dwelling predators. See Austin Hughes, "The Topography of Vision in Mammals of Contrasting Life Style: Comparative Optics and Retinal Organisation," Ch. 11 in Frederick Crescitelli, ed., *The Visual System in Vertebrates* (Berlin: Springer Verlag, 1977), pp. 613–756.
2. Boughey 1975, pp. 31–35 (listed among references for Ch. 2). Cf. Campbell 1974, pp. 84–90.
3. See Campbell 1974, pp. 196–202, 234–241; Washburn and Moore 1974, pp. 61–82.

4. See Ehrlich, Holm, and Brown 1976 (listed among references for Ch. 2), pp. 482–483.
5. Campbell 1974, pp. 384–401.
6. This idea was ingeniously developed by Cannon 1932.
7. Kraus 1964, pp. 285–288.
8. Compare with each other the following: Cannon 1932, pp. 287–306; Campbell 1974, pp. 401–408; Montagu 1965, pp. 199–213; Sears 1957, pp. 45–60; and Salk 1972, pp. 53–118.
9. Stebbins 1970.
10. Harry M. Caudill, "Farming and Mining: There Is No Land to Spare," *The Atlantic Monthly* 232 (Sept., 1973): 85–90.
11. Campbell 1974, pp. 241–244.
12. Cf. Dice 1955, pp. 28–30.
13. Cf. Sears 1957, p. 21.

Selected References

Campbell, Bernard G.
   1974. *Human Evolution.* 2nd ed. Chicago: Aldine.
Cannon, Walter B.
   1932. *The Wisdom of the Body.* New York: W. W. Norton.
Crick, Francis
   1966. *Of Molecules and Men.* Seattle: University of Washington Press.
Dice, Lee R.
   1955. *Man's Nature and Nature's Man: The Ecology of Human Communities.* Ann Arbor: University of Michigan Press.
Dubos, Rene
   1968. *So Human an Animal.* New York: Charles Scribner's Sons.
Kraus, Bertram S.
   1964. *The Basis of Human Evolution.* New York: Harper & Row.
Montagu, Ashley
   1965. *The Human Revolution.* New York: Bantam Books.
Salk, Jonas
   1972. *Man Unfolding.* New York: Harper & Row.
Sears, Paul B.
   1957. *The Ecology of Man.* Eugene: Oregon State System of Higher Education.
Stebbins, G. Ledyard
   1970. "The Natural History and Evolutionary Future of Mankind." *American Naturalist* 104 (Mar.-Apr.): 111–126.
Washburn, S. L., and Ruth Moore
   1974. *Ape into Man: A Study of Human Evolution.* Boston: Little, Brown.

10 | Industrialization: Prelude to Collapse

Unrecognized Preview

The Industrial Revolution made us precariously dependent on nature's dwindling legacy of non-renewable resources, even though we did not at first recognize this fact. Many major events of modern history were unforeseen results of actions taken with inadequate awareness of ecological mechanisms. Peoples and governments never intended some of the outcomes their actions would incur.

To see where we are now headed, when our destiny has departed so radically from our aspirations, we must examine some historic indices that point to the conclusion that even the concept of succession (as explored in previous chapters) understates the ultimate consequences of our own exuberance. We can begin by taking a fresh look at the Great Depression of the 1930s, an episode people saw largely in the shallower terms of economics and politics when they were living through it.[1] From an ecologically informed perspective, what else can we now see in it?

The Great Depression, looked at ecologically, was a preview of the fate toward which mankind has been drawn by the kinds of progress that have depended on consuming exhaustible resources. We need to see why it was not recognized for the preview it was; this will help us to grasp at last the meaning missed earlier.

We did not know we were watching a preview because, when the world economy fell apart in 1929–32, it was not from exhaustion of essential fuels or materials. From the very definition of carrying capacity—the maximum *indefinitely* supportable ecological load—we can now see that non-renewable resources provide *no* real carrying capacity; they provide only phantom carrying capacity. If coming to depend on phantom carrying capacity is a Faustian bargain that mortgages the future of *Homo colossus* as the price of an exuberant present, *that* mortgage was not yet being foreclosed in the Great Depression. Even so, much of the suffering that befell so much of mankind in the 1930s does need to be seen as the result of a carrying capacity deficit. The fact that the deficit did not stem from resource exhaustion in that instance makes it no less indicative of the kinds of grief entailed by resource depletion. Accordingly, we need to understand what did bring on a carrying capacity deficit in the 1930s.

Carrying Capacity and Liebig's Law

To attain such an understanding we need to step outside the usual economic or political frames of thought, go back two-thirds of a century before the 1929 crash, and reexamine for its profound human relevance a principle of agricultural chemistry formulated in 1863 by a German scientist, Justus von Liebig.[2] That principle set forth with great clarity the concept of the "limiting factor" briefly mentioned in Chapter 8. Carrying capacity is, as we saw there, limited not just by food supply, but potentially by *any* substance or circumstance that is indispensable but inadequate. The fundamental principle is this: whatever necessity is least abundantly available (relative to per capita requirements) sets an environment's carrying capacity.

While there is no way to repeal this principle, which is known as "the law of the minimum," or Liebig's law, there is a way to make its application less restrictive. People living in an environment where carrying capacity is limited by a shortage of one essential resource can develop exchange relationships with residents of another area that happens to be blessed with a surplus of that resource but happens to lack some other resource that is plentiful where the first one was scarce.

Trade does not repeal Liebig's law. Only by knowing Liebig's law, however, can we see clearly what trade does do, in ecological terms. Trade enlarges the scope of application of the law of the minimum. The composite carrying capacity of two or more areas with different resource configurations can be greater than the sum of their separate carrying capacities. Call this the principle of scope enlargement; it can be expressed in mathematical notation as follows:

$$CC_{(A + B)} > CC_A + CC_B.$$

The combined environment (A + B) still has finite carrying capacity, and that carrying capacity is still set by the necessary resource available in least (composite) abundance. But if the two environments are truly joined, by trade, then scarcities that are local to A *or* B no longer have to be limiting.

A good many of the events of human history need to be seen as efforts to implement the principle of scope enlargement. Most such events came about as results of decisions and activities by men who never heard of Liebig or his law of the minimum. Now, however, knowing the law, and understanding also the scope-enlargement principle, we can see important processes of history in a new light. Progress in transport technology, together with advancements in the organization of commerce, often achieved only after conquest or political consolidation, have had the effect of enlarging the world's human carrying capacity by enabling more and more local populations (or their lifestyles) to be limited not by local scarcity, but by abundance at a distance.

Vulnerability to Scope Reduction

As human numbers (and appetites) grew in response to this exchange-based enlargement of composite carrying capacity, continued access to non-local resources became increasingly vital to human well-being and survival. As the ecological load increased beyond what could have been supported by the sum of the separate carrying capacities of the formerly insulated local environments, mankind's vulnerability to any disruption of trade became more and more critical. The aftermath of the crash of 1929 demonstrated that vulnerability.

Unfortunately, modern transport systems, and some aspects of modern organization, are based very heavily upon exhaustible resource exploitation. Insofar as this is true, they must eventually founder upon the rocks of resource exhaustion. But even before they might succumb to such physical disaster, the trade arrangements upon which the earth's extended carrying capacity for *Homo colossus* has come to depend can be torn apart by *social* catastrophe.[3] It is important to recognize at last that that is what happened in 1929–32. In fact, some of it began happening during, or as a repercussion of, the Great War of 1914–18.

World War I disrupted relationships between the various peoples of Europe and between Europe, the New World, and the Orient. It also resulted in reallocation of the still colonial parts of the world among the various imperial powers seeking to exploit them as ghost acreage. Not all aspects of these changes wrought by the war would have reduced the scope of application of Liebig's law, but some certainly did, for some peoples, to some extent.

In the case of defeated Germany, access to resources from outside

German territory was cut off. At the same time, the staggering requirement of reparations payments to the victorious Allies aggravated the load to be borne by Germany's limited indigenous carrying capacity. Even internally, Germany suffered as inflation shattered the vital exchange relations between its diverse localities and between the occupational categories (quasi-species) into which its culturally advanced population had become differentiated.[4] Destruction of the value of currency meant destruction of the medium of mutualism; as inter-occupational symbiosis crumbled, hardship was rampant.

The astronomical German inflation was thus no mere fluke of history. Rather, it was a preview of the larger preview to come, when other forms of financial disruption would rend the fabric of trade throughout the world. By thus compelling a reduction of the scope of application of Liebig's law back down to local resource bases, such trade dislocation would convert existing loads of human resource-consumers, previously supportable by composite carrying capacity, into overloads no longer fully supportable by fragmented carrying capacities.

In America in the 1920s, after a brief post-war depression, a period of neo-exuberance set in, leading in the later years of the decade to such an *expectation* of perpetual progress and prosperity that some people found they could prosper from the expectation itself. "Speculation" in the stock market became the expected way to get rich.[5] Inhibitions against speculation were relaxed; people supposed the American prototype democracy, having enabled the Allies finally to triumph over Kaiser Germany, had made the world safe for getting rich and had established the right of everyone to try to do so.

The essential contrast between speculation and genuine investment is this: speculators buy stock not for the purpose of acquiring claims on future dividends from the business in which they acquire shares, but for the purpose of profiting from the expected escalation in their stock's resale value. When nearly all buyers are speculators, then virtually the only value of their shares is the resale value. Stock prices continue to escalate under such circumstances only as long as virtually everyone expects resale values to continue rising, and are thus willing to buy. The fact that prices may already grossly exaggerate a stock's intrinsic (dividend-paying) worth simply ceases to concern the speculator during the time when price escalation is confidently expected to continue. Breakdown of that faith, however, turns the process around. Anticipation of inexorable enrichment gives way to fear of ruin as self-induced price escalation turns into self-induced price decline. Panic, in the stock market sense, means the competitive drive to sell before falling prices fall farther—which drives prices down.

What connected the 1929 Wall Street crash to Liebig's law was the fact that so much speculative buying had been done with borrowed money. Collapse in the "value" of stocks thus led to an epidemic of bank failures, because the banks were unable to retrieve the funds they had lent to the speculators. Stock certificates taken in by the banks as security from borrowers were worth much less money after the crash than the number of dollars borrowed on them before the crash. When banks failed, depositors with accounts in those banks suddenly found themselves shorn of the purchasing power formerly signified by their bankbook entries. As depositors went broke, they ceased being able to buy goods or hire employees. Sellers of whatever they would have bought, or workers they would have employed, were therefore also suddenly bereft of revenue sources. In a society with elaborate division of labor and a money economy, a "revenue source" is the magic key that provides access to carrying capacity. Collapse of fiscal webs thus confronted millions of people with loss of access to carrying capacity, as truly as if purchasable resources had actually ceased to exist. Nations whose citizens had increasingly become masters of one trade apiece and jacks of few others found themselves suddenly unable to rely on composite carrying capacity drawn from a nonlocal environment. What I have called the "medium of mutualism" was no longer functioning, so the scope of application of Liebig's law of the minimum was being constricted once again to local (or personal) resources.

There was not in those days any Federal Deposit Insurance Corporation to back up the solvency of an individual bank when it suffered a "run" by its depositors. The failure of bank after bank in a time when banks had no institutionalized way of pooling their assets for mutual protection can thus be seen as a fiscal instance of the hazards of scope reduction. Had bankers understood that an ecological principle formulated by an agricultural chemist could apply to the world of finance, perhaps something like the FDIC would have been invented sooner.

The fiscal collapse had an even more important implication than this for our ecological understanding of the human predicament. That implication appears in the generalized Depression that followed. Consider the farm population in America. Like almost everyone else, farm families were compelled, by the repercussions of bank failures and the ramifications of general panic, to cut their consumer expenditures. Farmers also often had to allow their land, their buildings, and their equipment to deteriorate for lack of money to pay for maintenance and repairs. Many farms were encumbered by mortgages—mortgages which were foreclosed by banks that now desperately needed the payments farmers could not afford to make. (Bank failures were even more common in rural regions than in major cities.) In spite of all these difficulties, however, the farm population in America ceased declining (as it had been doing) and increased between 1929 and 1933 by more than a million. The long-term trend of movement out of farm niches and into urban niches was reversed during the Great Depression.[6]

Niches everywhere were being constricted by the Depression. However, the urbanizing trend that had been occurring as a result of industrial growth in the cities and from elimination of farm niches by mechanization of agriculture was disrupted by this economic breakdown. At the heart of the reversal was a simple fact: the nature of farming in the 1930s was still such that, whatever else they had to give up, there was still truth in the cliché that "the farm family can always eat." Other (non-food-producing) occupational groups that now had to fall back (like the farmers) on carrying capacities of reduced scope could find themselves in much more dire straits.

If we read it rightly, then, we can see the differential impact of the Depression upon farm versus non-farm populations as a cogent indicator of the dependence of the total population on previously achieved enlargements of the scope of application of Liebig's law. With breakdown of the mechanisms of exchange, various segments of a modern nation had to revert as best they could to living on carrying capacities again limited by locally least abundant resources, rather than extended by access to less scarce resources from elsewhere. Although scope reduction hurt everyone, rural folk had local resources to fall back upon; urban people, in contrast, had so detached themselves as to have almost ceased to recognize the indispensability of those resources. For reasons we shall examine in a moment, economic hard times hit the farms sooner than they hit the cities, but in the final scope-reducing crunch the farmers turned out to have an advantage sufficient to interrupt a clear trend of urbanization.

No Fairy Godmother

The Depression also interrupted the advance of industrialization and its attendant occupational diversification of the population. With hindsight, that interruption becomes an opportunity to bring the previous diversification into ecological focus.

An ecological perspective enables us to see pressure toward niche diversification as the natural result of the overfilling of existing niches. Among non-human organisms, this pressure leads eventually to the emergence of new species. Among humans it leads through sociocultural processes to the emergence of new occupations (quasi-species), which, as we noted in Chapter 6, had been made clear by Emile Durkheim as long ago as 1893. To bring Durkheim's analysis and the ecological perspective to bear upon the Great Depression, however, we must take into account the fact that nature is no Fairy Godmother and provides no guarantee that new niches will automatically be already available at the right time and in the right quantity to absorb immediately the surplus population from overfilled previous niches. Nor does nature guarantee pre-adaptation of the surplus individuals to whatever new niches do become available.

In nature, overfilling of old niches can result in massive death. Many organisms fall by the wayside in the march of speciation. Among *human* organisms the principles hold, but the process is moderated because humans are occupationally differentiated by social processes rather than by biological processes. Ostensibly, when old niches become obsolete, we can retrain ourselves for new roles. So, for *Homo sapiens,* overpopulation and death are *avoidable* results of niche saturation. The avoidance is not easy, however, and retraining for new niches can be traumatic.

An ecological perspective thus heightens the significance of a classic sociological study that clearly showed how unlikely it is, even among members of the relatively flexible and plastic human species, that re-adaptation to new niches (as old ones close up) will occur easily or automatically. Between 1908 and 1918, W. I. Thomas at the University of Chicago analyzed mountains of documentary data on the experience of Polish immigrants in America.[7] The people he studied had come to the New World after absorbing the folkways of their native Poland. In America they were faced with the necessity of adapting to unfamiliar circumstances. Thomas found that old ways of behaving and thinking were not easily abandoned or changed. New ways were learned only with difficulty when they contradicted the migrants' old-country upbringing. Thomas generalized from the immigrants' situation to say something about social change in broader contexts. He concluded that an accustomed way of behaving tends to persist as long as circumstances allow. When circumstances change, making familiar and comfortable ways unworkable (or unacceptable), a degree of crisis is inevitable. Re-adaptation hurts. It is resisted.[8]

We know now that the change that makes re-adaptation necessary need not be relocation. Any event that makes old ways unworkable and new ways mandatory can provoke the trauma of reorientation. Conflict and tension are natural accompaniments of change; they tend to continue until some new *modus vivendi* is worked out. The new form of adaptation will typically combine some elements of the old with some features imposed by the changed circumstances.

"Culture shock" became a familiar term for denoting the enervating disorientation and bewilderment associated with movement into unfamiliar societal contexts. Even a casual tourist can feel it when he travels abroad. Half a century after the phenomenon was studied by W. I. Thomas among Polish peasants resettled in America, Alvin Toffler coined and popularized another phrase that extended the concept. "Future shock" was his apt new term; forced adjustment to *new* ways can be as traumatic as forced adjustment to *foreign* ways.[9]

People in a post-exuberant world found themselves surrounded by alien conditions. They underwent a great deal of future shock, years before they got that name for it. By mechanization of agriculture in the nineteenth and early twentieth century, the Western world greatly reduced the number of farm workers needed to provide sustenance for themselves and for urban dwellers. Displaced from agricultural occupations, ex-farmers naturally migrated into cities in search of alternative employment, employment for which their farming experience or upbringing had not prepared them. Industrial expansion connected with World War I took up the slack temporarily, making employable on an emergency basis many persons who would otherwise have been passed over as unprepared for a given job. The war also helped hasten the mechanization of agriculture that was creating the displaced farm-worker surplus. After the war, urbanization and the proliferation of industrial occupations could not altogether keep pace with the continuing displacement of workers from the farming sector. There continued to be more farmers than were needed, so the agricultural portion of the economy was beset with "overproduction." This depressed farm prices—several years before the Wall Street crash provided the impetus that depressed prices for everyone. The resulting loss of purchasing power by the farming population helped depress, in turn, the urban-industrial sectors of the world's economy.

Ecological difficulties were aggravated, of course, by human errors—the glibly confident indulgence in speculation in 1928 being one example. But the causal importance of some human errors was easily overestimated. Amid the economic and political events of 1929–32 it was plausible for Americans, unaware of the ecological basis for what was happening, to see all the difficulties of that difficult time as products merely of the failures of the Hoover administration. This attractive oversimplification neglected one fact that should have been obvious: many other nations, over which Mr. Hoover did not preside, were undergoing the same calamity.

For those of radical inclination, it seemed plausible (in the absence of an ecological paradigm) to attribute the dire situation to a failure of "the capitalist system." But socialists believed as ardently as capitalists in the myth of limitlessness. In spite of socialists' commitment to production for use rather than for profit, they were not then (and have not been since) any more cautious than capitalists about adopting the drawdown method. They assumed that socialist-sponsored versions of drawdown could somehow eliminate such "capitalist contradictions" as simultaneous overproduction and abject poverty. They remained just as unconcerned as the capitalists about overshoot.[10]

Conservatives, on the other hand, who were not necessarily misanthropes, found it plausible to whistle in the dark, insisting that prosperity would automatically return if we just waited for the system to adjust itself. They were the Ostriches of their time, holders of the Type V attitude (delineated in Chapter 4). They believed nothing essential had changed from the Age of Exuberance.

Roosevelt was elected to replace Hoover, new approaches were put rapidly into practice, and a discouraged nation took heart. But full economic recovery continued to elude even the New Deal until preparation for World War II began to spur massive industrial activity—with even more than the usual disregard for long-range drawdown costs.

Economic recovery under the New Deal was not unique. Nazi Germany also overcame its depression, reducing unemployment in the first four years under Hitler from six million to one million. (People outside Germany did not automatically interpret this achievement as validation of Nazi tactics.) Under the Nazi method, millions of the unemployed could be employed as soldiers, and millions more could be *compulsorily* retrained and given niches as producers of military hardware. The war economy nurtured demand for consumer goods for the soldiers and for these re-employed makers of military materiel; furthermore, it provided "the correct psychological atmosphere," enabling the civilian sector to accept painful re-adaptation.

War psychology overcame natural human resistance to departure from custom.[11] The war also used elaborate technology and drew down the world's stocks of natural resources.

In the United States, wartime economic recovery supposedly proved that New Deal "pump priming" by fiscal deficits had been the right kind of response to a stagnant economy, except that it could not be done in adequate volume until the need to re-arm rapidly for all-out war made truly massive red-ink budgets politically acceptable. But American recovery from the depression of the 1930s did not unambiguously validate the Keynesian economic theory implicit in Roosevelt's approach.

In either the German or the American portion of the Great Depression, an economic interpretation (by minds unaccustomed to an ecological perspective) enabled us to miss the point. Very simply, the ecological paradigm enables these events to be read as follows: Expansion of the military establishment, at the cost of additional resource drawdown, suddenly provided new niches (in industry and in the armed forces) capable of absorbing the overflow from the whole array of saturated civilian occupations. And the wartime social climate provided the patriotic push that made the trauma of re-adaptation to new occupational roles endurable. The new or enlarged military-industrial niches had been previously either non-existent or under serious stigma. What was important, ecologically speaking, was the fact that previously existent and acceptable niches *had been saturated;* there were people to spare—in America because of technological progress and population growth; in Germany because of the debacle of World War I and its aftermath, which left the German economy, occupational structure, and national morale in a shambles. Moreover, human redundancy throughout much of the world had become manifest when, in various ways and in various places, the medium of mutualism came apart, leaving everyone to cope with carrying capacity limits set by local minimums.

In the American case, the fiscal deficits run up during World War II were merely the ledger-book picture of the change that eased the problem, not the cause of that change. Red ink didn't re-employ the unemployed. The growing national debt (expressed in money) was a fiction of accountancy, a fiction that enabled Americans to believe that wartime drawdown of the once–New World's resource reservoir only constituted "borrowing from ourselves," rather than stealing from the future. The reality of diachronic competition remained unacknowledged. Nevertheless, resources used up in World War II were made unavailable for use by posterity.

Circular versus Linear Ecosystems

Whatever the origins of human redundancy, and whatever the sequel to it, we needed to see (but were not seeing) that what had happened to us between the wars, and especially what happened to us since World War II, had not resulted merely from politics or economics in the conventional sense. The events of this period had simply accelerated a fate that began to overtake us centuries ago. The population explosion after 1945 and the explosive increase of technology during and after the war were only the most recent means of that acceleration.

Human communities once relied almost entirely on organic sources of energy—plant fuels and animal musclepower—supplemented very modestly by the equally renewable energy of moving air and flowing water. All of these energy sources were derived from ongoing solar income. As long as man's activities were based on them, this was, as church men said, "world without end." That phrase should never have been construed to mean "world without limit," for supplies can be perpetual without being infinite.

Locally, green pastures might become overgrazed, and still waters might be overused. Local environmental changes through the centuries might compel human communities to migrate. As long as resources available *somewhere* were sufficient to sustain the human population then in existence, the implication of Liebig's law was that carrying capacity (globally) had not yet been overshot. If man was then living within the earth's current income, it was not from wisdom, but from ignorance of the buried treasure yet to be discovered.

Then the earth's savings, and new ways to use them, began to be discovered. Mankind became committed to the fatal error of supposing that life could thenceforth be lived on a scale and at a pace commensurate with the rate at which treasure was discovered and unearthed. Drawing down stocks of exhaustible resources would not have seemed significantly different from drawing upon carrying capacity imports, at a time when nobody yet knew Liebig's law, or the principle of scope enlargement, or the distinction between real and phantom carrying capacity, or the various categories of ghost acreage.

Homo sapiens mistook the rate of withdrawal of savings deposits for a rise in income. No regard for the total size of the legacy, or for the rate at which nature might still be storing carbon away, seemed necessary. *Homo sapiens* set about becoming *Homo colossus* without wondering if the transformation would have to be quite temporary. (Later, our pre-ecological misunderstanding of what was being done

167

to our future was epitomized by that venerable loophole in the corporate tax laws of the United States, the oil depletion allowance. This measure permitted oil "producers" to offset their taxable revenues by a generous percentage, on the pretext that their earnings reflected depletion of "their" crude oil reserves. Even though nature, not the oil companies, had put the oil into the earth, this tax write-off was rationalized as an incentive to "production." Since "production" really meant *extraction,* this was like running a bank with rules that called for paying interest on each withdrawal of savings, rather than on the principal left in the bank. It was, in short, a government subsidy for stealing from the future.)

The essence of the drawdown method is this: man began to spend nature's legacy as if it were income. Temporarily this made possible a dramatic increase in the quantity of energy per capita per year by which *Homo colossus* could do the things he wanted to do. This increase led, among other things, to reduced manpower requirements in agriculture. It also led to the development of many new occupational niches for increasingly diversified human beings. (Expansion of niches in Germany, America, and elsewhere from 1933 to 1945 was, it now appears, just a brief episode in this long-run development.) Because the new niches depended on spending the withdrawn savings, they were niches in what amounted to a "detritus ecosystem." Detritus, or an accumulation of dead organic matter, is nature's own version of ghost acreage.[12]

Detritus ecosystems are not uncommon. When nutrients from decaying autumn leaves on land are carried by runoff from melting snows into a pond, their consumption by algae in the pond may be checked until springtime by the low winter temperatures that keep the algae from growing. When warm weather arrives, the inflow of nutrients may already be largely complete for the year. The algal population, unable to plan ahead, explodes in the halcyon days of spring in an irruption or bloom that soon exhausts the finite legacy of sustenance materials. This algal Age of Exuberance lasts only a few weeks. Long before the seasonal cycle can bring in more detritus, there is a massive die-off of these innocently incautious and exuberant organisms. Their "age of overpopulation" is very brief, and its sequel is swift and inescapable.

When the fossil fuel legacy upon which *Homo colossus* was going to thrive for a time became seriously depleted, the human niches based on burning that legacy would collapse, just as detritovore niches collapse when the detritus is exhausted. For humans, the social ramifications of that collapse were unpleasant to contemplate. The Great Depression was, as we have seen, a mild preview. Detritus ecosystems flourish and collapse because they lack the life-sustaining biogeochemical circularity of other kinds of ecosystems. They are nature's own version of communities that prosper briefly by the drawdown method.

The phrase "detritus ecosystem" was, of course, not widely familiar. The fact that "bloom" and "crash" cycles were common among organisms that depend on exhaustible accumulations of dead organic matter for their sustenance was not widely known. It is therefore understandable that people welcomed ways of becoming colossal, not recognizing as a kind of detritus the transformed organic remains called "fossil fuels," and not noticing that *Homo colossus* was in fact a detritovore, subject to the risk of crashing as a consequence of blooming.

Bloom and crash constitute a special kind of sere; certain kinds of populations in certain kinds of circumstances typically experience these two seral stages—irruption followed by die-off. Crash can be thought of as an abrupt instance of "succession with no apparent successor." As in ordinary succession, the biotic community has changed its habitat by using it, and has become (much) less viable in the changed environment. If, after the crash, the environment can recover from the resource depletion inflicted by an irrupting species, then a new increase of numbers may occur and make that species "its own successor." Hence there are *cycles* of irruption and die-off (among species as different as rodents, insects, algae). Our own species' uniqueness cannot be counted upon as protection. Moreover, some of the resources we use cannot recover.[13]

When yeast cells are introduced into a wine vat, as noted in Chapter 6, they find their "New World" (the moist, sugar-laden fruit mash) abundantly endowed with the resources they need for exuberant growth. But as their population responds explosively to this magnificent circumstance, the accumulation of their own fermentation products makes life increasingly difficult—and, if we indulge in a little anthropomorphic thinking about their plight, miserable. Eventually, the microscopic inhabitants of this artificially prepared detritus ecosystem all die. To be anthropomorphic again, the coroner's reports would have to say that they died of self-inflicted pollution: the fermentation products.

Nature treated human beings as winemakers treat the yeast cells, by endowing our world (especially Europe's New World) with abundant but exhaustible resources. People promptly responded to this circumstance as the yeast cells respond to the conditions they find when put into the wine vat.

When the earth's deposits of fossil fuels and mineral resources were being laid down, *Homo sapiens* had not yet been prepared by evolution to take advantage of them. As soon as technology made it possible for mankind to do so, people eagerly (and without foreseeing the ultimate consequences) shifted to a high-energy way of life. Man became, in effect, a detritivore, *Homo colossus*. Our species bloomed, and now we must expect crash (of some sort) as the natural sequel. What form our crash may take remains to be considered in the concluding section.

One thing that kept us from seeing all this, and enabled us to rush exuberantly into niches that had to be temporary, was our ability to give ideological legitimation to occupations that made no sense ecologically. When General Eisenhower, as retiring president, warned the American people to beware of unwarranted influence wielded by the military-industrial complex,[14] it was presumably political and economic influence that he had in mind. But the military-industrial complex was a vast conglomeration of occupational niches. As such, it wielded an altogether different (and even more insidious) kind of influence. The military-industrial complex helped perpetuate the illusion that we still had a carrying capacity surplus; it made it profitable for the living generation to extract and use up natural resources that might otherwise have been left for posterity. It absorbed for a while most of the excess labor force displaced by technological progress from older occupational niches that had been less dependent on drawing down reservoirs of exhaustible resources. It thus helped us believe that the Age of Exuberance could go on.

Nor was General Eisenhower alone in missing the ecological significance and over-emphasizing the political elements in the trends of his time. His young, articulate, and sophisticated Bostonian successor launched a new administration with an inaugural address whose inspirational quality lay partly in its eloquent resolution of American ambivalence. If we wanted to maintain full employment, we dreaded achieving it by means of an arms race. Subtly, and with the gloss of high idealism, John F. Kennedy reassured the nationwide television audience on that crisp, brilliant January day in 1961 that the temporary occupational niches of the military-industrial complex could be long-lasting and could be made more honorable than horrible. There was to be a "new Alliance for Progress," and we were to hope for emancipation from the "uncertain balance of terror that stays the hand of mankind's final war." But the conflict-bred niches would last, for "the trumpet summons us again . . . to bear the burden of a long twilight struggle year in and year out . . . against the common enemies of man: tyranny, poverty, disease and war itself."[15]

Under both parties, the military-industrial complex enabled us to be preoccupied with matters that helped us ignore resource limits. It helped thereby to obscure the fact that population was expanding to fill niches that could not be permanent because they were founded upon drawing down prehistoric savings, exhaustible fossil energy stocks.

The human family, even if it were soon to stop growing, had committed itself to living beyond its means. *Homo sapiens,* as we saw in Chapter 9, was capable of transforming himself into new "quasi-species." By the Industrial Revolution humans had turned themselves into "detritovores," dependent on ravenous consumption of long-since accumulated organic remains, especially petroleum.

If we were to understand what was now happening to us and to our world, we had to learn to see recent history as a crescendo of human prodigality. When American birth rates declined as the 1960s gave way to the 1970s, this did not mean we were escaping the predicament of the algae any more than the ringing words of President Kennedy's inaugural address had really meant that we could eat our cake and still have it. Rather, something had happened that was fundamental, and that could not be undone by brilliant rhetoric: there had been a marked acceleration in our previously begun shift from a self-perpetuating way of life that relied on the circularity of natural biogeochemical processes, to a way of life that was ultimately self-terminating because it relied on linear chemical transformations. They were linear (and one way) because man was using (with the aid of his prosthetic equipment) so many non-crop substances. Man was no longer engaged in a balanced system of symbiotic relations with other species. When man degraded the habitat, it tended to stay degraded; it was not being rehabilitated by other organisms with different biochemical needs.

Perils of Prodigality: The Coming Crash

Man does not live on detritus alone. Misled by our prodigal expenditures of savings, we allowed the human family to multiply so much that by the 1970s mankind had taken over for human use about *one-eighth* of the annual total net production of organic matter by *contem-*

porary photosynthesis in all the vegetation on all the earth's land. That much was being used by man and his domestic animals.[16] It would require taking over more than the other seven-eighths to provide from organic sources the vast quantities of energy we were deriving from fossil fuels to run our mechanized civilization, even if economic growth and human increase were halted by the year 2000. Thus, as we began to see in Chapter 3, we were already well beyond the size that would permit us to re-adapt (without severe depopulation) to a sustained yield way of life when our access to savings gave out. On the other hand, just three more doublings of population (scarcely more than Britain had already experienced in the short time since Malthus) would mean that *all* the net photosynthetic production on all the continents and all the islands on earth would have to be used for supporting the human community. Then our descendants would be condemned to living at an abjectly "underdeveloped" level, if no fossil acreage remained available to sustain modern industry.

Such total exploitation of an ecosystem by one dominant species has seldom happened, except among species which bloom and crash. Detritovores provide clear examples, but there are others, and we shall take a close look at some of them in the final chapter. For *Homo sapiens*, it was unlikely that we could even divert much more than the already unprecedented fraction of the total photosynthesis to our uses.

It was thus becoming apparent that nature must, in the not far distant future, institute bankruptcy proceedings against industrial civilization, and perhaps against the standing crop of human flesh, just as nature had done many times to other detritus-consuming species following their exuberant expansion in response to the savings deposits their ecosystems had accumulated before they got the opportunity to begin the drawdown.

It was not widely recognized, of course, but the imminence of that kind of culmination really was why the United Nations had to convene its 1972 Conference on the Human Environment. The conference in Stockholm was meant to begin the process of preventing our only earth from being rendered less and less usable by humans. In short, its purpose was to arrest global succession. Persons who had struggled valiantly to bring about this conference had been engaged (in an important sense) in a global counterpart of the efforts of Dr. Goodwin in Williamsburg. But whereas he sought to undo succession in order to preserve history, they sought to preserve a world ecosystem in which *Homo sapiens* might remain the dominant species—and might remain human.

Until the extent of the transformation of *Homo sapiens* into *Homo colossus* was seen and the full ecological ramifications of that transformation were more nearly understood, however, it would hardly be recognized that the kind of world ecosystem the United Nations was seeking to perpetuate was already being superseded—by an ecosystem that, by its very nature, compelled the dominant species to go on sawing off the limb on which it was sitting. Having become a species of superdetritovores, mankind was destined not merely for succession, but for crash.

Unfortunately but inevitably, the Stockholm deliberations were confused by the fact that the luckier nations which happened to achieve industrial prodigality before the earth's savings became depleted had already infected the other nations with an insatiable desire to emulate that prodigality. The infection preceded recognition of the depletion. The result of this sad historical sequence was the pathetic quarrel over whether the luxury we cannot afford is economic growth or environmental preservation. Neither was a luxury; worse, neither was possible on a global scale.

Excess numbers and ravenous technology had already brought *Homo colossus* to an ecological impasse. The laudable ability of delegations from 114 diverse nations to hammer out compromise resolutions favoring both environmental protection *and* economic development for all nations did not extricate us from our predicament. Deft avoidance of political deadlock once again preserved the illusion that cake could be both eaten and saved. But illusion preserved was still illusion.

Man needed to realize how commonly populations of other species have undergone the experience of resource bankruptcy. But we humans have been experiencing a double irruption, confronting us with an intensified version of the plight of such species. As a biological type, *Homo sapiens* has been irrupting for 10,000 years, and especially the last 400. In addition, our detritus-consuming tools have been irrupting for the last 200 years. It is conceivable that the inevitable die-off necessitated by overshoot could apply more to *Homo colossus* than to *Homo sapiens*. That is, resource demand might be brought back within the limits of permanent carrying capacity by shrinking ourselves to less colossal stature—by giving up a lot of our prosthetic apparatus and the high style of living it has made possible. This might seem, in principle, an alternative to the more literal form of die-off, an abrupt increase in human mortality. In practice, it runs afoul of several implications of W. I. Thomas's finding about resistance to change. Accustomed ways of behaving and thinking tend to persist; this is probably as true of the detritovorous habits of *Homo colossus* as it was true of earlier human folkways. Outbreaks of violence among American motorists waiting in long queues to buy gasoline, sputtering in stubborn non-recognition of the onset of the twilight of the petroleum era, suggest that the people of industrial societies who have learned to live in colossal fashion will not easily relinquish their seven-league boots, their heated homes, and their habit of living high on the food chain. As we said, re-adaptation hurts. It will be resisted.

Moreover, habits of *thought* persist. As we shall see in Chapter 11, people continue to advocate further technological breakthroughs as the supposedly sure cure for carrying capacity deficits. The very idea that technology caused overshoot, and that it made us too colossal to endure, remains alien to too many minds for "de-colossalization" to be a really feasible alternative to literal die-off. There is a persistent drive to apply remedies that aggravate the problem.

If any substantial fraction of the more colossal segments of humanity *did* conscientiously give up part of their resource-devouring extensions out of humane concern for their less colossal brethren, there is no guarantee that this would avert die-off. It might only postpone it, permitting human numbers to continue increasing a bit longer, or less colossal peoples to become a bit more colossal, before we crash all the more resoundingly.

All this tends to be disregarded by advocates of a "return to the simple life" as a gentle way out of the human predicament. Blessed are the less prosthetic, for they shall inherit the ravaged earth. Probably so, in the long run. But some view the dark cloud of fuel depletion and purport to see a silver lining already: individuals forced to abandon much of their modern technology will then get by on smaller per capita shares of the phantom carrying capacity upon which prosthetic man has become so dependent. However, insofar as the high agricultural yields upon which our irrupted population's life depends can be attained only by means of energy subsidies—by lavish application of synthetic fertilizers, and by large-scale use of petroleum-powered machinery—the dwindling fossil acreage will probably lower the output of visible acreage. As we asked before, what happens when it becomes necessary again to pull the plow with a team of horses instead of a tractor, and a substantial fraction of the crop acreage that now feeds humans has to be allocated again to growing feed for draft animals (or biomass to produce tractor fuel when the Carboniferous legacy is no longer cheaply available)? So much for *that* silver lining.

It will spare us no grief to deny that *Homo sapiens* has been irrupting. It will in no way ease the impact to deny that crash must follow. We must seek our rays of hope in another way altogether (as we shall do in Chapter 15).

Not Cleared for Takeoff

The "developed" nations have been widely regarded as previews of the future condition of the "underdeveloped" countries. It would have been more accurate to reverse the picture, as perhaps the Stockholm Conference began to do for its most perceptive participants and observers.

It was one thing to be an underdeveloped nation in the eighteenth century, when the world had no highly developed nations. It is quite another thing today. When today's developed nations were not yet industrialized and were just approaching their takeoff point, the *World* had only recently entered an exuberant phase which made takeoff possible. European technology was just starting to harness (for a few brilliant centuries) the energy stored in the earth during the past several hundred million years, and the sparsely populated New World had only recently become available for exuberant settlement and exploitation. These conditions of exuberance no longer prevail. The underdeveloped countries of Asia, Africa, and Latin America in the twentieth century cannot realistically expect to follow in the footsteps of the undeveloped nations of eighteenth-century Europe. *Most of today's underdeveloped nations are destined never to become developed.* Egalitarian traditions will be forced to adjust to permanent inequality.

Hard as it might be for the people and leaders of underdeveloped countries to face the fact, they are not alone in finding it repugnant. The people and leaders of the affluent societies have also resisted seeing it. Recognition that most of the world's poor would necessarily stay poor would destroy the comforting conviction of the world's privileged that their good fortune ought to inspire the world's poor to emulate them, not resent them.

Nature's limiting factors would not clear most underdeveloped countries for takeoff. But now that people are so numerous, it would be even worse if many did somehow take off. Most men of good will have been unable so far to accept this implication of the ecological facts. Some will no doubt righteously denounce this book for analyzing the situation in this unpalatable way, as if no fact could hurt us if we refused to acknowledge its truth. But not only are there not enough of the substances a developed human community must *take from* its environment in the process of living to permit a world of four billion people to be all developed; the capacity of the world's oceans, continents, and atmosphere to absorb the substances *Homo colossus* must *put* somewhere in the process of living is limited. Even as a waste disposal site, the world is finite.

Right into the 1970s we were misled by so bland a word as "pollution" for this part of our predicament. We were already suffering the plight of the yeast cells in the wine vat. Accumulation of the noxious and toxic extrametabolites of high-energy industrial civilization had become a world problem, but no government could admit that it would turn into a world disaster if the benefits of modern technology were bestowed as abundantly upon everyone in the underdeveloped countries as they already had been upon the average inhabitant of the overdeveloped ones. Leaders everywhere had to pretend full development of the whole world was their ultimate aim and was still on the agenda. By such pretensions mankind remained locked into stealing from the future.

Learning to Read the News

Viewing contemporary events from a pre-ecological paradigm, we missed their significance. From an ecological paradigm we can see that fewer members of the species *Homo colossus* than of the species *Homo sapiens* can be supported by a finite world. The more colossal we become, the greater the difference. What we called "pollution," and regarded at first as either a mere nuisance or an indication of the insensitivity of industrial people to esthetic values, can now be recognized as a signal from the ecosystem. If we had learned to call it "habitat damage," we might have read it as a sign of the danger inherent in becoming colossal. Even if the world were not already overloaded by four billion members of the species *Homo sapiens*, it does not have room for that many consumers of resources and exuders of extrametabolites on the scale of modern *Homo colossus*. In short, on a planet no larger than ours, four billion human beings simply cannot all turn into prosthetic giants.

As we move deeper into the post-exuberant age, one of the keen insights of a passionately concerned and unusually popular sociologist, C. Wright Mills, will become increasingly important to us all. It was an insight by which he tried to help his contemporaries read the news of their times perceptively. We will need to be at least as perceptive to avoid misconstruing events that will happen in the years to come.

Although the paradigm from which Mills wrote was pre-ecological, in one of his most earnest books he transcended archaic thought-ways enough to note that only sometimes and in some places do men *make* history; in other times and places, the minutiae of everyday life can add up to mere "fate." Mills gave us an unusually clear definition of this important word. Infinitesimal actions, if they are numerous and cumulative, can become enormously consequential. Fate, he explained, is shaping history when *what happens to us was intended by no one and was the summary outcome of innumerable small decisions about other matters by innumerable people.*[17]

In a world that will not accommodate four billion of us if we all become colossal, it is both futile and dangerous to indulge in resentment, as we shall be sorely tempted to do, blaming some person or group whom we suppose must have intended whatever is happening to happen. If we find ourselves beset with circumstances we wish were vastly different, we need to keep in mind that to a very large extent they have come about because of things that were hopefully and innocently done in the past by almost everyone in general, and not just by anyone in particular. If we single out supposed perpetrators of our predicament, resort to anger, and attempt to retaliate, the unforeseen outcomes of our indignant acts will compound fate.

In precisely Mills's sense, the conversion of a marvelous carrying capacity surplus into a competition-aggravating and crash-inflicting deficit was a matter of fate. No compact group of leaders ever decided knowingly to take incautious advantage of enlargment of the scope of applicability of Liebig's law, or subsequently to reduce that scope and leave a swollen load inadequately supported. No one decided deliberately to terminate the Age of Exuberance. No group of leaders conspired knowingly to turn us into detritovores. Using the ecological paradigm to think about human history, we can see instead that the end of exuberance was the summary result of all our separate and innocent decisions to have a baby, to trade a horse for a tractor, to avoid illness by getting vaccinated, to move from a farm to a city, to live in a heated home, to buy a family automobile and not depend on public transit, to specialize, exchange, and thereby prosper.

Notes

1. See the explanations offered by various analysts cited in Patterson 1965, pp. 227–245.
2. For the original formulation of this principle, see Liebig 1863, p. 207. Also see the sharpened statement of it on p. 5 in the "Editor's Preface" to that volume. For indications that Liebig had the principle in mind even before he grasped its generality and fundamental significance, see his earlier work, *Chemistry in Its Application to Agriculture and Physiology* (London: Taylor & Walton, 1842), pp. 41, 43, 85, 127, 129, 130, 132, 139, 141–142, 159, 178. On the development of Liebig's thinking about this and other ecological principles, see Justus von Liebig, "An Autobiographical Sketch," trans. J. Campbell Brown, *Chemical News* 63 (June 5 and 12, 1891): 265–267, 276–278; W. A. Shenstone, *Justus von Liebig: His Life and Work* (New York: Macmillan, 1895); and Forest Ray Moulton, ed., *Liebig and After Liebig: A Century of Progress in Agricultural Chemistry* (Washington: American Association for the Advancement of Science, 1942).
3. Cf. Fred Hirsch, *Social Limits to Growth* (Cambridge: Harvard University Press, 1976). Too often social limits are unwisely cited as if to afford some basis for disregarding environmental finiteness; social limits actually make finiteness all the more salient. They do not make carrying capacity less relevant to human affairs. The cliché which asserts "There are no real shortages, only maldistribution" inverts the significance of social limits. In comparison with biogeochemical limits, social limits to growth include all the ways in which human societies are prone to fall short of developing and maintaining the optimum organization that would allow Liebig's law to apply only on a thoroughly global scale, with carrying capacity thus never limited by *local* shortages. Social limits, in other words, tend to aggravate, not alleviate, the problems posed by biogeochemical limits.
4. See William L. Shirer, *The Rise and Fall of the Third Reich* (New York: Simon and Schuster, 1960), pp. 61–62. In thinking about the human implications of the law of the minimum and the social impediments to implementing the principle of scope enlargement, it is well to remember that, when the collapse occurred in Germany, one ramification was the opportunity it afforded for rise of the Nazi dictatorship, with grave consequences for many other nations.
5. See Galbraith 1955, especially the first five chapters.
6. See Ch. 4, "Farmers in the Depression," in Chandler 1970.
7. See Thomas and Znaniecki 1918–1920 passim.
8. Cf. Robert A. Nisbet, *Social Change and History* (New York: Oxford University Press, 1969), pp. 282–284.
9. Toffler 1970, pp. 4–5.

10. Cf. Ehrenfeld 1978 (listed among references for Ch. 1), pp. 249–254. For recent examples of socialist persistence in the myth of limitlessness, see Stanley Aronowitz, *Food, Shelter and the American Dream* (New York: Seabury Press, 1974); Hugh Stretton, *Capitalism, Socialism and the Environment* (New York: Cambridge University Press, 1976). Also see Irving Louis Horowitz, *Three Worlds of Development: The Theory and Practice of International Stratification*, 2nd ed. (New York: Oxford University Press, 1972), p. xvi, where "overdevelopment" is defined without any ecological reference as "an excess ratio of industrial capacity to social utility," i.e., to the ability of *people* with existing organization, skill levels, etc., to benefit from industrial output. In contrast, overdevelopment signifies to ecologists—e.g., Ehrlich and Ehrlich 1972 (listed among references for Ch. 12), pp. 418–420—a level of technological development that disregards *physical and biological limitations* and requires "far too large a slice of the world's resources to maintain our way of life."

11. Michael Tanzer, *The Sick Society* (New York: Holt, Rinehart and Winston, 1971).

12. See, for example, Odum and de la Cruz 1963; Darnell 1967.

13. This makes it unwise to have defined these substances as "resources."

14. For an interesting discussion of the political significance of Eisenhower's warning, see Fred Cook, *The Warfare State* (New York: Macmillan, 1962).

15. Quoted and discussed in Morison 1965 (listed among references for Ch. 5), p. 1110.

16. Odum 1971 (listed among references for Ch. 6), p. 55.

17. Mills 1958, pp. 10–14.

Selected References

Chandler, Lester V.
   1970. *America's Greatest Depression 1929–1941.* New York: Harper & Row.
Commoner, Barry
   1971. *The Closing Circle: Nature, Man, and Technology.* New York: Alfred A. Knopf.
Darnell, Rezneat M.
   1967. "The Organic Detritus Problem." Pp. 374–375 in George H. Lauff, ed., *Estuaries.* Washington: American Association for the Advancement of Science, Publication no. 83.
Galbraith, John Kenneth
   1955. *The Great Crash 1929.* Boston: Houghton Mifflin.
Hubbert, M. King
   1969. "Energy Resources." Ch. 8 in Committee on Resources and Man, *Resources and Man.* San Francisco: W. H. Freeman.

Jensen, W. G.

    1970. *Energy and the Economy of Nations.* Henley-on-Thames, Oxfordshire: G. T. Foulis.

Liebig, Justus

    1863. *The Natural Laws of Husbandry.* New York: D. Appleton.

Mills, C. Wright

    1958. *The Causes of World War Three.* New York: Simon and Schuster.

Odum, Eugene P., and Armando A. de la Cruz

    1963. "Detritus as a Major Component of Ecosystems." *American Institute of Biological Sciences Bulletin* 13 (June): 39–40.

Odum, Howard T.

    1971. *Environment, Power, and Society.* New York: John Wiley & Sons.

Patterson, Robert T.

    1965. *The Great Boom and Panic 1921–1929.* Chicago: Henry Regnery.

Thomas, William Isaac, and Florian Znaniecki

    1918–1920. *The Polish Peasant in Europe and America.* 5 vols. Chicago: University of Chicago Press; Boston: Richard Badger.

Toffler, Alvin

    1970. *Future Shock.* New York: Random House.

Watson, Adam, ed.

    1970. *Animal Populations in Relation to Their Food Resources.* Oxford: Blackwell.

V | Resistance and Change

... any social hope that is going to be any use against the darkness ahead will have to be based upon a knowledge of the worst: the worst of the practical facts, the worst in ourselves.
— C. P. Snow
The State of Siege, pp. 19–20

In one of the most perceptive and imaginative sentences in *Das Kapital,* Marx wrote: "The country that is more developed industrially only shows to the less developed, the image of its own future." It is significant to note that what the mid-nineteenth century must have considered a wildly romantic thought is a commonplace today.
— Irving Louis Horowitz
Three Worlds of Development: The Theory and Practice of International Stratification, p. 3.

The peoples of Asia, Africa, and Latin America are also demanding the fruits of industrialization. But to give every one of the ... people on earth the American standard of living would put an unbearable drain on the world's finite resources. ... To raise per capita energy consumption to the U.S. level, the world would need to burn 300 percent more coal, 500 percent more petroleum, and 1100 percent more natural gas.
— Thomas J. Kimball
"Status of the Environment and Our Efforts to Improve It," *Proceedings of the 1972 International Conference on Nuclear Solutions to World Energy Problems,* p. 6

Such a drain on the exhaustible resources of the earth would deplete many of them in a generation or two; but as the demand for them increased with the progressive industrialization, their value would certainly skyrocket, and the current international stresses ... would undoubtedly become excessive
— Richard T. LaPiere
Social Change, p. 540

11 | Faith versus Fact

Vain Expectations

As an allegorical representation of one of the common reactions to our post-exuberant situation, consider the plight of an elderly cancer patient who came to understand that her doctors could not cure her affliction. Her response was to take up a religious faith that denied the reality of bodily ailments by defining them as signs of spiritual weakness. It taught her to expect spiritual devotion to bring about a miracle which medicine could not perform.

But her cancer continued. As death approached, she was then doubly tormented—by physical pain no less than before, but now also by the anguish of guilt, for the undeniable evidence of advancing disease became a basis for self-reproach, being a sign (she now supposed) of the insufficiency of her religious fidelity.

Americans in an age of overshoot came to suffer as this woman suffered. The New World was growing old, and it pained us to see the accumulating social and political effects of the closing of America's frontier, to feel the depletion of the earth's savings, to endure the technological degradation of land, sea, and air, and to be reviled from time to time as "greedy" by peoples we had supposed would just naturally learn to emulate our "progressive" ways. For many, these pains were aggravated by a needless sense of shame, for people imagined that, despite all the transformations described in Chapters 7, 8, 9, and 10, only a loss of national will-power prevented us from revitalizing and universalizing the American dream. When the ecological basis for traditional expectations had ceased to exist, however, fulfillment of the dream was simply no longer possible. In post-exuberant circumstances, even such will-power as had once seemed invincible *had to be* insufficient.

When circumstances by mid-1979 had raised to a critical level of urgency the world's need for an ecologically enlightened United States energy policy, the American president concluded that he must address his countrymen not simply with specific energy-management proposals, but also in terms of deeper difficulties reflected in the nation's inability to unite on energy matters. He discerned a crisis of confidence. "We've always believed," he said, "in something called progress. We've always had a faith that the days of our children would be better than our own." But Americans, he correctly noted, "are losing that faith."[1]

Because that faith had become obsolete, we had to lose it. No previous generation of Americans had to cope with cumulative effects of their own and their ancestors' thefts from the future, as ours must. But instead of wisely discussing with his listeners how and why the traditional faith had become inapplicable, the president sought to revive it. "We can regain our confidence," he insisted, and earnestly invoked as inspiration toward that end a national heritage from generations who, he mistakenly asserted, had "survived threats much more powerful and awesome than those that challenge us now."

Accordingly, some of the remedies he went on to propose for the energy predicament were of a Cargoist nature. He called for "the most massive peacetime commitment of funds and resources in our nation's history," unmindful for the moment that past overcommitment of resources not perpetually available underlay the impasse the nation now faced. He called for creation of "an Energy Security Corporation" to lead a national effort "to replace two and a half million barrels of imported oil per day by 1990" with alternative fuels. With a revised target year, this was essentially a revival of Project Independence from the Nixon administration.

Several months later this unrealism was compounded when Edward Kennedy announced his candidacy for the presidential office. Seeming to imply that no problem could have yet become insoluble, he misconstrued Carter's July speech as an instance of "blaming" the people for their malaise, and proposed by his own style of "leadership" to carry forward the American faith in limitlessness. Other persons competing for high office also let themselves be tempted into scoffing at the "malaise" idea, as if erosion of faith in a utopian future either had not happened or should not.

The Millenarian Response

Faced with mounting indications of the inability of technological or political efforts to prolong the American dream, many people in social strata and age brackets which once would have shunned religious fantasies turned to them in the 1960s and 1970s to revive dashed hopes. New adherents were attracted to movements that ranged from (1) quiet renewal of interest in conventional religious observance, through (2) Pentecostal episodes within the usually sedate denominations (e.g., Episcopalians "speaking in tongues"), to (3) strongly millenarian fads—the "Jesus freaks," the Satanists, assorted devotees of non-Western doctrines, and even practitioners of occult beliefs and rituals.

Movements proclaiming the imminence of the millennium and calling upon believers to prepare for it tend to arise among, and appeal to, people who earnestly seek release from some felt oppression.[2] In the past, colonial peoples had been receptive to millenarian beliefs. So had the subordinated or dispossessed members of feudal societies. Expectation of the millennium has often led to actions that aggravated the believers' actual plight—causing expectant stoppage of economic activity, or ritual destruction of resources.

Chronic dissatisfaction and yearning breed millenarian cults. People need not have suffered actual material deprivation; heightened desires can produce equivalent dissatisfaction. The neo-exuberant "revolution of rising expectations," together with the deterioration of the worldwide ecological basis for fulfilling such expanding hopes, have tended to foster millenarian beliefs and activities.

As indicated in Chapter 4, our understanding of such developments in the modern context can be enlarged by taking a comparative look at the cargo cults in the Pacific island societies studied extensively by anthropologists.[3] To the pre-literate Melanesian peoples in the Pacific islands, European society was unseen and baffling. The Europeans who came to the islands brought strange ways and imported many material things. The processes by which these goods had been produced, the type of social organization and equipment which enabled them to be produced in such quantity and variety remained unseen and unknown. It was apparent to the islanders that European people had some secret magic; they obtained abundant cargoes of material objects without laboring to create them. The Europeans seen in the islands by the Melanesians did not make things and did not do menial tasks; they got the natives to do menial work for them. So it was not implausible for the natives to suppose that any laborious production processes in Europe or America must have been carried out by laborers like unto the Melanesians—perhaps even their own dead ancestors.

When the natives learned to want the sorts of things Europeans had, and tried to imagine how the Europeans came to possess them, it was easy to infer that working for the Europeans was not the only way, or the most justified way, of obtaining European cargo. The idea that the cargo brought in by ships and delivered to the Europeans in the islands had been stolen from black people in Europe seemed, to the Melanesians, consistent with their observation of the relatively work-free lives of white men. And if the cargo had been stolen from their kinsmen, natives could rightfully claim it. To obtain it, they might only have to discover and practice the proper magic or ritual.

Modern faith in science and technology as infallible solvers of any conceivable problem can be, in a post-exuberant world, just as superstitious. The essential parallel is this: the Melanesians were able to believe they would receive cargo because they had no accurate knowledge of how European goods came into existence, or why they came to the islands. The modern Cargoist who expects to be bailed out of this year's ecological predicament by next year's technological breakthrough holds similar beliefs because of his inadequate knowledge of ecology and of technology's role in it. Both Cargoist faiths rest upon the quicksand of fundamental ignorance lubricated by superficial knowledge.

On the basis of such beliefs, the inability to satisfy inflated wants often led the Melanesian people into hysterical and paranoid behavior (trances, twitching, mass possession). It sometimes led to destructive acts and avoidance of work. But it also led to meticulous construction of wharfs to receive the expected ships. Warehouses were built to store the anticipated cargo. By working diligently to facilitate delivery of cargo, Melanesians affirmed their belief that it was rightfully theirs.

America is not Fiji, the Admiralty Islands, or New Guinea. But Americans, too, have been subjected to the equivalent of an "invasion." *Homo colossus* has overfilled niches once sparsely (and thus comfortably) occupied by *Homo sapiens*. Except for an occasional perceptive scholar like Sumner or Turner (or, before them, the rare political intellect like Jefferson), Americans until recently remained as unaware of the ecological basis for their dream and their accomplishments as the Melanesians were of the European factory system. They had little or no comprehension of ghost acreage, or of the geological processes that had made available to them the stored energy of ancient photosynthesis. In believing that democratic political doctrines and the magic of free enterprise had sufficed to bring forth the good life yesterday, or in supposing that steadfast ideological faith would retrieve it today, some Americans were embracing their own equivalent of a cargo cult—for similar reasons. Each, in his own way, was clinging to the myth of limitlessness.

Space Age Cargo Cults

The cults that won Cargoist adherents among the citizens of advanced nations were not always obviously religious. The Type II belief held that great technological breakthroughs would inevitably occur in the near future, and would enable man to continue indefinitely expanding the world's human carrying capacity.[4] This was a mere faith in a faith, like stock-market speculation; it had no firmer basis than naive statistical extrapolation—the uncritical supposition that past technological advances could be taken as representative samples of an inherently unending series of comparable achievements. Such a faith overlooked the fact that man's ostensible "enlargement" of the world's productivity in the past had mainly consisted of successive *diversions* of the world's life-supporting processes from use by other species to use by man. It failed to see that "progress" (even by the takeover method) must stop when all divertable resources have been diverted. Man obviously can't take over more than everything. (Less obviously, there are biological and geological reasons why considerably less than 100 percent of the world's resources could be diverted to human use.)

Technological optimism manifested itself in several pious hopes, enumerated below.

1. "Unlimited" food

Enthusiasm for the Green Revolution was merely a special case of this cult of great technological breakthroughs. It believed that the crucial breakthrough had already been achieved (by development of high-yield strains of wheat and rice), and that now vigorous missionary effort throughout the hungry nations would convince their peoples to raise these superior crop varieties.[5] Such an attitude was just another expression of inability to understand, or reluctance to perceive, the finiteness of the biosphere. Believers in this "breakthrough" were unable to see that further extension of the human irruption was going to be a problem aggravated, not a problem solved. They could not even see that the high-yield grains either would hasten the exhaustion of the soils on which they were grown or would intensify agriculture's precarious dependence upon a chemical fertilizer industry. Cultivation of renewable resources such as food was becoming heavily dependent upon continuing depletion of exhaustible resources like petroleum and minerals.[6] Man's efforts to enlarge carrying capacity by agricultural progress were thus making the old takeover method dependent upon the treacherous drawdown method.

187

2. *"Unlimited" alternatives*

In response to the worldwide shortfall of energy supplies under burgeoning demand, economists extolled "resource substitution" as the answer. The man in the street tended to believe that somehow "new sources" of energy would be tapped, to make us "self-sufficient" by some vague target date—1980 at first, then 1990, then. . . . These were some of the millenarian pipe-dreams that arose to assuage post-exuberant anxieties. American oil companies in the early part of 1974 bid up to $7,000 per acre to lease oil-shale "development" rights;[7] if crude oil prices climbed high enough, these companies would start devouring mountains to extract shale oil. Their bids and intentions thus lent a peculiar aptness to the phrase of an English writer who wondered how many people "can safely play the planet-eating game."[8]

In Chapter 1 it was pointed out that two non-repeatable achievements had made possible four centuries of magnificent progress. Those two achievements were (1) the discovery of a second hemisphere, and (2) development of technology that could unearth and exploit the planet's energy savings, its fossil fuel deposits. Mankind's increasingly relentless search for new sources of energy and for more costly energy technologies expresses our wish to deny that achievements like those two were uniquely resultant from bygone circumstances.

3. *"Unlimited" energy*

In earlier times occasional dedicated eccentrics who pipe-dreamed of violating the laws of physics tried to invent "perpetual motion" machines—to provide energy, supposedly, without consuming any fuel. These impossible devices had always seemed like an ideal means to perpetuate limitlessness. Now, in what we had thought were more sophisticated times, Cargoists permitted themselves to believe in the same sort of absurdity by talking glibly of the "breeder reactor" as a device that not only would generate vast quantities of energy but also would, in the process, "produce more fuel than it consumes." It would *not* do that, of course. The illusion that it would, and therefore that mankind could expect to continue treating the world as limitless, arose from careless phrasing that fostered misunderstanding of rather simple physical facts.

The fuel for nuclear fission power plants was the heavy element uranium. More than 99 percent of the uranium in the world was the heavier isotope, U-238, which would not enter into a chain reaction and generate power. The much scarcer U-235 would do so. Producing fuel for a reactor (or an atomic bomb) at first consisted of "enriching" natural uranium—that is, sorting out the fissionable U-235 atoms from the useless but more abundant U-238 atoms. However, if a neutron hurtling out of a fission reaction was captured by a U-238 nucleus before being slowed down by collisions with other atoms, it could turn the U-238 atom into Plutonium-239, which was fissionable. The so-called breeder reactor, then, was a device to enable some of the energy released by fission to be fed back to convert the non-fuel U-238 into the fuel Pu-239. It was not a device that could make fuel out of nothing, any more than was an oil refinery (which also refines more fuel than it burns). The breeder reactor would merely enable man to reduce the unusable fraction of the available uranium. The illusion of limitlessness was sustained, then, not by the device itself, but by its name and the sloppiness of its common description. These devices did not offer "unlimited" quantities of energy for human use. They would multiply the usable energy content of known uranium reserves by a factor of up to sixty—not by infinity, or even by some astronomical number.[9]

Besides, there was the question of safety.[10] To be operationally safe, fast-breeder reactors would have to have at least the following features: (1) provision for infallibly shutting down the fission reaction under any circumstances that might arise; (2) means of assuring uninterrupted flow of coolant, or means of at least detecting with certainty any flow interruptions while they were still only incipient—i.e., before they permit dangerous overheating to occur; (3) means of preventing any escape of fission products in case of fuel melting as a result of failure of either of the above systems; (4) means of preventing discharge of any radioactive effluent; (5) facilities for collection and very long-term storage of dangerously radioactive substances that are unavoidably created as by-products of the energy-generating process. Cargoism either overlooked the difficult engineering and organizational problems implicit in these safety requirements or glibly assumed they were soluble.

Cargoists were fond, too, of dreaming about nuclear *fusion* as the ultimate source of "limitless" energy.[11] Deuterium, a heavy isotope of hydrogen, would serve as fuel in a reaction that could occur only at temperatures of millions of degrees—such as are found naturally in the interior of the sun. Handling the ionized gases involved in the fusion process imposed sufficient technical difficulties so that *expecting* energy problems to be solved by fusion was clearly a case of counting upon uncertain-to-hatch chickens.[12] These unearthly plasmas must somehow be confined, for dissipation would instantly drop their temperature. Since no material container could be kept from vaporizing at even a fraction of the high temperature required, it quickly came to be assumed that containment must be accomplished by very strong magnetic fields. To generate such intense magnetism required electromagnets involving superconductive components. Materials can be made superconductive by cooling them down nearly to absolute zero. Thus, at the outset, the whole idea of earthly nuclear fusion depended on assuming it would be feasible to engineer devices that could achieve in close proximity temperatures that were both unearthly high and unearthly low.

Fusion advocates inverted Liebig: preoccupation with the *abundance* of the hydrogen that would serve as fuel blocked perception among these Cargoists of a *scarcity or absence* of structural materials that would not become brittle at extreme temperatures and under exposure to intense radiation. Such materials would need to retain strength, maintain size and shape, resist fatigue, and not blister, sputter, or erode. Expectations of high energy outputs have prevented lay enthusiasts for thermonuclear power from considering such problems as the possibility that intense magnetic fields might alter the heat-conducting properties of the liquid metals that would be expected to serve as coolants. There were also enormous problems in the matter of adapting electrical generating facilities to use the intense bursts of heat energy at very high temperatures emanating from thermonuclear reactions.

Popular mythology supposed that fusion power would be altogether free from problems of radioactivity, since hydrogen rather than uranium or plutonium would serve as the fuel. One system of fusion would use tritium (the heavier isotope of hydrogen with atomic weight = 3) as fuel; tritium was radioactive and could leak by diffusion through hot metal walls. Even the system that would use deuterium as the fuel would produce some tritium in the process, and would thus require coping with problems of possible tritium leakage. Moreover, as research toward fusion power progressed, it began to appear that the most economically feasible system was going to be one that was a kind of hybrid of fission *and* fusion, a system that would accept hazards of radioactive waste disposal as the necessary price for simply making it feasible to begin using the cheap and abundant hydrogen as an auxilliary fuel.[13]

The non-physicists who were counting thermonuclear ergs many years before an exceedingly problematic hatching were indulging in faith based on a little knowledge and much ignorance—just as Melanesians' expectations of cargo rested upon knowing very little and assuming too much about the nature of European production processes.

4. *"Harnessing" the sun*

The ultimate fall-back position of the modern Cargoist was the expectation that new technology would eventually "enable us to use solar energy."[14] This view overlooked the ways in which man was already heavily dependent upon solar energy.

In more ways than one, solar energy supported the agriculture that had enabled *Homo sapiens* to irrupt from a few million inhabitants of the earth in pre-Neolithic times to some five hundred times as many only 400 human generations later. Solar energy supported agriculture not only through photosynthesis; it also supplied the energy for evaporation which was "pumping" each day some 68.6 trillion gallons (= 260 cubic kilometers) of water from the surfaces of land and sea up into the atmosphere, whence it could rain down upon the world's farms, forests, and hydroelectric watersheds.[15]

If only $\frac{1}{10}$ of 1 percent of the solar energy that reached the earth's surface was captured by plants and fixed in organic molecules, this did not mean the other 99.9 percent was a "vast untapped reservoir" awaiting man's exploitation. It could be exceedingly dangerous for mankind to try using even an additional 0.1 percent; the difference between an untapped 99.9 percent and an untapped 99.8 percent might seem trivial, but it would be an imposition upon the energy system of the ecosphere comparable to that already being made by the entire standing crop of organisms of all kinds.

The Cal Tech geochemistry professor Harrison Brown suggested back in 1954 that, a century hence, a world population of seven billion people could conceivably be living at an "American" level of energy use, and might be deriving one-fourth of that energy from solar devices.[16] Rather simple calculations will show, however, that this would entail diverting to human use an amount of solar energy roughly three times as great as the entire quantity of energy used by the world's population in the year Brown made the suggestion. To put this in perspective, consider the fact that the total human use of energy is already equivalent to more than 10 percent of the total net organic production by the entire biosphere. To supply future humans with three times that much from solar devices means doing *something* to the largely unknown natural pattern of energy flow on a scale that is not infinitesimal after all. *Homo colossus* would be swinging almost as much weight as a third of the whole biosphere! The potentially disruptive effects upon the balanced processes of nature have to constitute an enormous risk.

5. *Other technological escapes*

Cargoism was not always so flagrantly recognizable as in these examples. Had the Solomon Islanders been Europeans rather than Melanesians, perhaps instead of laboriously constructing neat rows of cargo storage houses they would have invested their hopes in a project like the British-French Concorde (to bring on the millennium by expensively providing the blessings of supersonic travel for the world's businessmen and diplomats). No less exaggerated were the hopes invested by France in her unilateral program for achieving nuclear weapons parity.

The belief that emigration by space ship to unpopulated worlds would exempt us from the consequences of overshooting this world's carrying capacity flourished briefly when the Space Age was beginning, but it was no more realistic than postwar construction of "airstrips" by Melanesians. (The latter had updated their cultist expectations of cargo delivery methods after American military aviation came into their lives during World War II.) Believers in extraterrestrial emigration as a solution to irrupting population never seemed to do the simple arithmetic to estimate the prodigious tonnage of space vehicles and impossible quantitites of fuel it would take to boost up to escape velocity the earth's yearly increment of population (some 70 million human beings), plus the supplies they would need on their long journey to some *hypothetically inhabitable* other planet. We are, of course, talking of exporting 70 million people per year for the rest of their lives, never to return. In the entire Apollo program of manned lunar landings, five pairs of astronauts spent a total of just over 23 man-days on the moon. If we could hope to hold the line on earthly population by exporting our growth to a planet no more inaccessible than the moon (as Europe once exported surplus people to the New World), it would take more than 60,000 Apollo-type launchings *every day* to do it! Even if that were not absurdly infeasible, scientific space research had quickly undercut any illusions of escape by this means, for the two most plausible planetary destinations, Venus and Mars, were shown by space probe photography and telemetry to be impossible environments for massive colonization by human refugees from the post-exuberant earth.

6. *Ideological escapes*

Other radical enthusiasms were non-technological in content. Just as the people of the Pacific islands supposed they could obtain, by appropriate religious activities, vast quantities of cargo whose industrial basis was unknown to them, so the assorted revolutionaries in the industrialized world from 1963 onward imagined that drastic redistribution of status and influence, adoption of new "lifestyles," relaxation of social restraints, and creation of a new "love ethic" could bring on the human benefits of a noncompetitive era more sublime than the imperfect but inspiring one that had been rather effectively nurtured by the carrying capacity surplus the world once enjoyed. Revolutionaries remained oblivious to what a bulletin of the Yale University School of Forestry astutely called "the ecological limits of optimism."

A case in point was the popularity of Charles Reich's book, *The Greening of America*. This popularity must be credited to the earnest desire of young people living in the twilight of exuberance for reassurance that simple acquisition of a new view of the world would, in effect, repeal certain laws of nature and exempt us from the consequences of the irruption that had been happening since Neolithic times. Reich and his eager readers seemed unaware that the increasingly competitive relations among members of the human species were the natural consequence of our inexorably changing ecological circumstances. Reich's views were a peculiar blend of Cargoism and Cosmeticism.[17] (He extolled the revolutionary implications of flared trousers, for example.) Except for the belief that drugs held promise of a shortcut to the millennium, Reich's "greening" was not particularly technological. He would have had us believe that we could transcend competitiveness by rejecting *doctrines* perpetrated by the corporate state. But replacing them with the "radical subjectivity" he advocated could not block the sequence of bloom and crash in which we had involved ourselves, any more than Melanesian rituals could bring European cargo to the islanders.

Reich's prescription for what he saw ailing us said we should abandon values artificially instilled by the corporate state, values that gave undue importance to status and power. We must "start from the premises based on human life and the rest of nature." But Reich's "nature" was an idol: nothing he said about it showed any awareness of the impact of resource limits, biogeochemical processes, symbiosis, ecological antagonism, etc., upon human relations. If he had any comprehension of the perils of overshoot, this was not evident in his expectation that mere graduation to what he called "Consciousness III" could free men altogether from their competitive predicament.[18] "Consciousness III" was thus a thoroughly millenarian illusion.

Looking for Scapegoats

The various forms of Cargoism all had in common a stubborn insistence that carrying capacity limits could be raised again, as they had been several times in the past. Many versions of Cargoism assumed technology was the means for doing this; some versions imagined it could be done just by changing our hearts and minds. Cargoists, especially of the latter persuasion, refused to believe that problems arising from overshoot were real and ineluctable. They therefore tended to resent Realists who sought to reveal the facts of a post-exuberant world. Cargoists seemed to say, in effect, "If we just don't accept them as facts, they won't *be* facts."

To the resistive mind, no new fact-revealing paradigm was acceptable unless it was palatable—i.e., unless its adoption would put things right as well as reveal the nature of our predicament. People with Cargoist outlooks were tempted, quite understandably, to suppose the world's troubles were not so much due to post-exuberant conditions as to spokesmen for a new paradigm who *pointed out* the conditions. According to Reich, for example, we needed to become deeply suspicious of rationality, logic, analysis, and principles. Defiance of previous authority has been a common component of cargo cults. It has had various modes of expression, such as the burning of sacred objects, the exposure of secret things to categories of persons for whom they had been taboo, or destruction of money and material possessions.

Had we known what acts to consider symptomatic, we might have seen Cargoist thoughtways and millenarian passions behind such modern events as the theft and publication of secret documents, the burning of flags or embassies, and perhaps even the vandalistic attack on a priceless Michelangelo sculpture in the Vatican, or the scratching of the letters I-R-A into a treasured painting in the King's College chapel at Cambridge. Some of the more sordid features of the American and European "counterculture" in the 1960s and 1970s can be viewed as post-exuberant instances of the sacrilege and ritual obscenity that have occurred in other times and places when the people of a despairing society took the antinomian path and tried to attribute their plight to a moral code that could be overthrown, rather than to circumstances that could not be escaped.

With restraint and inquiry, the circumstances might be *understood*. But understanding them might not enable us to change them, and people who could not abide the thought that our post-exuberant condition was due to real circumstances insisted that culprits must be causing it. If no obvious power-hungry tyrant or plutocrat was readily available for casting in the culprit role, there tended to be further insistence that the culprits were the ecologically awakened persons who made it their business to discover and point out the real circumstances. For example, one politically radical Protestant clergyman, writing in vigorous opposition to the "seduction of radicalism" by the "ecology movement" in the 1970s, compared such a movement's efforts to transcend politics in the name of nature to the German Nazi ideology of the 1930s. Hitler, he said, had urged Germans to live by the dictates of nature—i.e., give vent to their racial instincts.[19] (Apparently we must all forever ignore "nature" because Hitler misused the word!) In the name of compassion, writers like this clergyman deplored efforts to gain recognition for ecosystem constraints, as if the *fact* that the world was finite would not condemn burgeoning millions to a brutish existence, but *speaking the words* "limited carrying capacity" would bring on the horrors that must not happen.

Notes

1. For the text of the Carter speech, see *New York Times,* July 17, 1979, p. A15.
2. See Burridge 1969; Cohn 1970.
3. As examples of the literature on cargo cults, see Worsley 1957; Lawrence 1964. For a readable and illuminating review of interpretive analyses of these cults, see Jarvie 1963.
4. For examples of technological optimist writings, see Brown 1954; Seaborg and Corliss 1971; Weinberg 1973; Hawley 1975, p. 10. For criticism of technological optimism, see Udall et al. 1974; Hardin 1972, pp. 141–151.
5. Johnson 1972. On risks discerned in the Green Revolution itself and in reliance upon it, see Graham Chedd, "Hidden Peril of the Green Revolution," *New Scientist,* Oct. 22, 1970, pp. 171–173; Michael Allaby, "Green Revolution: Social Boomerang," *The Ecologist* 1 (Sept. 1970): 18–21.
6. See Perelman 1972; Steinhart and Steinhart 1974.

7. James P. Sterba, "210–Million Total Bid for Oil Shale," *New York Times,* Jan. 9, 1974, p. 15.

8. Johnson 1973.

9. Bennet 1972, pp. 59–64; Seaborg and Corliss 1971, pp. 34–43.

10. Lovett 1974; see also Philip M. Boffey, "Reactor Safety: Congress Hears Critics of Rasmussen Report," *Science* 192 (June 25, 1976): 1312–13; David White, "Nuclear Power: A Special New Society Survey," *New Society* 39 (Mar. 31, 1977): 647–650. As an indication of the impact of the nuclear accident in the Three Mile Island power plant on views of nuclear safety, see Carolyn D. Lewis, "Taking Nuclear Risks," *Newsweek,* Nov. 12, 1979, pp. 32–33; Victor K. McElheny, "Two Faces of Technology," *Science 80* 1 (Nov./Dec., 1979): 17–18.

11. Philip H. Abelson, "Glamorous Nuclear Fusion," *Science* 193 (July 23, 1976): 279; also see Hunt 1974; Seaborg and Corliss 1971, pp. 43–50, 105.

12. Some of the technical difficulties are discussed in William D. Metz, "Fusion Research (II): Detailed Reactor Studies Identify More Problems," *Science* 193 (July 2, 1976): 38–40, 76.

13. See William D. Metz, "Fusion Research (III): New Interest in Fusion-Assisted Breeders," *Science* 193 (July 23, 1976): 307–309.

14. As a premise, Cargoists of this type could well quote Daniels 1967, p. 15: "There is an ample supply of solar radiation falling on the earth to do all the work that will conceivably be needed, and a new supply will always be available every sunny day—as long as there are people on the earth." For examples of solar energy expectancy, see Meinel and Meinel 1976; Cheremisinoff and Regino 1978; Williams 1978.

15. If mean annual precipitation is taken to be a little over ½ meter over the ⅓ of the earth's surface that is land, it is easy to calculate from knowledge of the earth's radius and the geometric formula for the area of a sphere that the total rainfall volume on land would be roughly 260 cubic kilometers per day.

16. Brown 1954, pp. 184–186. In fairness to Brown, it should be noted that his technological optimism was tempered by concern that "a substantial fraction of humanity today is behaving as if . . . it were engaged in a contest to test nature's willingness to support humanity and, if it had its way, it would not rest content until the earth is covered completely and to a considerable depth with a writhing mass of human beings" (Brown 1954, p. 221). See also Brown 1978.

17. See, in particular, Reich 1970, Ch. 9.

18. In fact, Reich (1970, p. 166) naively declared that "the world is ample for all."

19. Neuhaus 1971, p. 157; see also pp. 86–87, where "the ecology movement" (rather than either the advertising agency or the oil company) is put down because an Amoco TV commercial had the effrontery to depict the purchase of lead-free gasoline as a revolutionary answer to the question of "What can one man do?" about air pollution.

Selected References

Bennet, D. J.
1972. *The Elements of Nuclear Power*. London: Longman.
Brown, Harrison
1954. *The Challenge of Man's Future*. New York: Viking Press.
1978. *The Human Future Revisited: The World Predicament and Possible Solutions*. New York: W. W. Norton.
Burridge, Kenelm
1969. *New Heaven, New Earth: A Study of Millenarian Activities*. New York: Schocken Books.
Cheremisinoff, Paul N., and Thomas C. Regino
1978. *Principles and Applications of Solar Energy*. Ann Arbor: Ann Arbor Science Publishers.
Cohn, Norman
1970. *The Pursuit of the Millennium: Revolutionary Millenarians and Mystical Anarchists of the Middle Ages*. New York: Oxford University Press.
Daniels, Farrington
1967. "Direct Use of the Sun's Energy." *American Scientist* 55 (Mar.): 15–47.
Hardin, Garrett
1972. *Exploring New Ethics for Survival: The Voyage of the Spaceship Beagle*. New York: Viking Press.
Hawley, Amos H., ed.
1975. *Man and Environment*. New York: New Viewpoints/*New York Times*.
Hunt, Stanley Ernest
1974. *Fission, Fusion and the Energy Crisis*. Oxford: Pergamon Press.
Jarvie, I. C.
1963. "Theories of Cargo Cults: A Critical Analysis." *Oceania* 34 (Sept.): 1–31; (Dec.): 109–136.
Johnson, Brian
1973. "The Planet-Eating Game." *People* 1 (Oct.): 7–9.
Johnson, Stanley
1972. *The Green Revolution*. London: Hamish Hamilton.
Lawrence, Peter
1964. *Road Belong Cargo: A Study of the Cargo Movement in the Southern Madang District, New Guinea*. Manchester: Manchester University Press.
Lovett, James E.
1974. *Nuclear Materials: Accountability Management Safeguards*. Hinsdale, Ill.: American Nuclear Society.
Meinel, Aden B., and Marjorie P. Meinel
1976. *Applied Solar Energy: An Introduction*. Reading, Mass.: Addison-Wesley.
Neuhaus, Richard
1971. *In Defense of People: Ecology and the Seduction of Radicalism*. New York: Macmillan.

Perelman, Michael J.
1972. "Farming with Petroleum." *Environment* 14 (Oct.): 8–13.
Reich, Charles
1970. *The Greening of America*. New York: Random House.
Seaborg, Glenn T., and William R. Corliss
1971. *Man and Atom: Building a New World through Nuclear Technology*. New York: E. P. Dutton.
Steinhart, John S., and Carol E. Steinhart
1974. "Energy Use in the U.S. Food System." *Science* 184 (Apr. 19): 307–316.
Udall, Stewart L., Charles Conconi, and David Osterhout
1974. *The Energy Balloon*. New York: McGraw-Hill.
Weinberg, Alvin M.
1973. "Technology and Ecology—Is There a Need for Confrontation?" *BioScience* 23 (Jan.): 41–45.
Williams, Robert H., ed.
1978. *Toward a Solar Civilization*. Cambridge: MIT Press.
Worsley, Peter
1957. *The Trumpet Shall Sound: A Study of 'Cargo' Cults in Melanesia*. London: MacGibbon & Kee.

12 | Life under Pressure

Attaining Perspective

A common expression of resistance to the new ecological paradigm was the act of pinning the label "neo-Malthusian" on any contention that the world was not limitless—as if applying a label to a valid piece of knowledge would refute it. Before anyone ever had occasion to try to stigmatize any of my ideas with that label, I was influenced in writing early drafts of some chapters of this book by the marvelously nostalgic experience of residing for several years in New Zealand, a young and hopeful nation that resembled in many important ways the United States in its exuberant phase, before it became an Old World nation.

When New Zealanders came home to their spacious and beautiful islands from travel overseas, their families and friends sometimes spoke of it as returning "from the world." Their enormously varied country seemed blessed in many ways by its remoteness from other lands and from the bulk of the planet's population.[1] But it was not really a separate world. It was on the same globe, was subject to the same laws of nature, and had abundant commerce with other nations. Its European settlers had introduced enough species of plants and animals from elsewhere in the world so that it was no longer the separate ecosystem it had been when the kiwi and other flightless bird species evolved.

New Zealanders were avid readers, and copious users of mass communications from overseas. Involved with the rest of mankind, they were protected somewhat, by location and low density, from humanity's most vexing problems. Looking at the world's troubles from their vantage point for several years helped me grasp the new ecological paradigm. Good-willed New Zealanders, I noticed, sometimes insisted that their government should do as much as possible to facilitate immigration, so as to relieve problems of overpopulation elsewhere on this planet. As a newcomer from a much larger land, I was sensitive to the finiteness of New Zealand's carrying capacity; at the same time, I was deeply appreciative of its still low density of settlement. Thus I was led to recognize that if a mere three weeks' addition to the world's population were all resettled in New Zealand, the population density of those lovely islands would then match the world average. Like the rest of the world, that country would have passed beyond its Age of Exuberance.

I came therefore to realize how lucky the United States was for having been an "emerging nation" at a time when the Old World's population pressure was temporarily relieved by the carrying capacity surplus of a New World. Underdeveloped countries in the closing decades of the twentieth century did not have the opportunity for vast expansion of man's ecological niches that had been provided three centuries earlier by colonization of a newly available hemisphere.[2] Alaska (and the arctic portions of Canada) now offered only an illusory equivalent to the vast land masses of the New World that had seemed so virgin when European explorers discovered them. These northern lands were large in area but small in carrying capacity because of climate. They could not absorb anything like the number of emigrants that had gone to the New World from an overpopulated Europe, or might now want to leave present areas of overpopulation. Even arid Australia had a carrying capacity so patently limited that its ability to perform the pressure-relieving function of a New World to the rest of the planet was almost as illusory, because it was so evidently temporary.

These remaining unpressured regions cannot significantly prolong the world's Age of Exuberance. But they still afford opportunities for a first-hand contemporary glimpse of that age to a few of the hundreds of millions of people who have already been exposed to post-exuberant life.

From the New Zealand perspective, it began to be apparent to me that termination of America's accustomed material progress might be less unwelcome to many Americans than they would generally have expected. Material progress had actually been contributing to the increasing mutual interference in each other's lives. Many people were finding this mutual interference so disturbing that they might appreciate an end to the escalation of affluence almost as much as they might dread it.

I began to realize how little the average American had at first understood the reasons for his own anguish over what was happening to the American way of life in the post-exuberant era. This could be brought out clearly, I discovered, by asking an American to name an overpopulated country. He or she was much more likely to begin by mentioning India than the United States. But in considering which nation was overpopulated to a greater extent, population *pressure* needed to be distinguished from population *density*.

Pressure and Density

Other things being equal, greater density would create greater pressure. But pressure and density are not the same, and other things were not equal.

Population pressure can be defined as the frequency of mutual interference per capita per day that results from the presence of others in a finite habitat. (This is about what Emile Durkheim meant by "moral density" in his 1883 analysis of the division of labor in society.[3]) Population density in the ordinary sense (Durkheim's "material density") is simply the number of people per square mile. Two nations with equal population density could differ in population pressure if their peoples differed in level of activity. A population using more prosthetic equipment would tend to subject its members to more pressure by doing more things.

As a model for assessing a country's population *pressure,* consider what can happen to a quantity of air in a closed container (such as an automobile tire). You can increase the pressure either by pumping in more air or by heating what is already there. If you heat it, you make each molecule more active. With greater energy, the molecules move about faster and collide more often. Thus the movement of each one is more often "interfered with" by the movements of others—just as if there were *more* others. Pressure thus depends on activity as well as on numbers.

In America, population had increased: more people had been pumped into our finite living space, to make demands upon our finite resources. But our pace of living had also been greatly accelerated. We traditionally welcomed such acceleration as a sign of progress, seldom recognizing that it meant people had increased the ways in which their co-presence resulted in mutual interference. The loss of independence and the failure to understand how it was lost can be illustrated by a fundamental change in the occupational structure of an industrialized nation's labor force. For example, in 1940 roughly 9 million Americans worked at farming occupations. By 1970, only about 2.5 million were so engaged.[4] In one standard sociology textbook, this change was shown on a graph, with a line sloping downward from left to right to represent the declining number of farm workers. Superimposed on the same graph is a line sloping steeply upward from left to right, representing the increasing number of "persons fed per farm worker"—implying that the change is necessarily a magnificent accomplishment. But the upward-sloping line could just as well have been labeled "number of consumers upon whom each farm worker's prosperity depends." In short, increased per capita productivity in farming has subjected farm workers to increased dependence upon others.

This trend had been running for a long time. In 1926 the dean of the graduate school at North Carolina State College, a rural sociologist named Carl C. Taylor, wrote:

> The division of labor . . . has been as beneficial to the farmer and his family as to anyone else. It has left him free to specialize in the production of raw materials and this specialization in no small way accounts for his increased efficiency. His increased efficiency in turn has made it possible for him to sell his products in the world markets and with the money received for them to buy more of the world's goods than he could ever have enjoyed under a system in which he supplied all his own and his family's needs out of his own fields, flocks, and herds. *To say that he is now specializing* in the production of raw goods *is but another way of saying he is depending on other people* to furnish him with finished goods. He is more efficient under this system but less self-sufficient.[5]

In 1973 the world was abruptly confronted with this linkage between increased efficiency and reduced self-sufficiency when the dependence of the "productive" industrialized nations upon the comparatively "underdeveloped" oil-exporting nations was used by some of the latter as a foreign policy lever.

Man's chief advantage, his capacity for quasi-speciation within one species, was fraught with a serious disadvantage, the intensification of precarious interdependence. The seriousness of that disadvantage was greatest for those portions of mankind that had learned to value independence strongly by living through an Age of Exuberance, or inheriting from it a culture of exuberance. Increased mutual interference, an integral feature of *modern* life, was all the more burdensome to people with cherished expectations of freedom to do as they pleased.

High-Energy Living

One measure of the change in our pace of life that had magnified our interference with each other was the increase in one generation in the per capita use of energy by Americans. From 1945 to 1970, for example, each American's use of fuel rose (on the average) 116 percent.[6] Per capita use of natural gas increased 287 percent. The electrical energy generated for each American increased 379 percent. Coal consumption per capita increased less than 13 percent as people shifted to more convenient energy sources.

Increased energy use was reflected in increased use of material resources. The generation that saw human population increase 45 percent (between 1940 and 1965) saw factory sales of automobiles increase 150 percent; on a per capita basis, annual car sales had risen 72 percent. One result was the sprawling automobile graveyards so common all across the United States but so conspicuously absent from other countries where the trade-it-in-after-two-years compulsion had not developed. Burgeoning automobile production in America had to mean either more cars on the road or more rapid depreciation in market value and earlier junking of "old" but still usable cars. Either way, progress entailed interference as well as assistance in people's pursuit of happiness.

American production of raw steel per person increased 15 percent between 1945 and 1970; primary production of aluminum per capita increased about 500 percent. Drawdown was clearly being accelerated.

In the past, we too easily thought of changes like these as a welcome growth of abundance. We overlooked the fact that even a fixed number of people living in a fixed area have to experience greater population pressure the more they turn themselves into a species of *Homo colossus*. Less prosthetic members of the species *Homo sapiens*, at the same density, were under less pressure. Increased automobile production per capita, for example, either had to mean increased traffic congestion (and driver frustration) or increased freeway construction (and community disruption, plus tax increases). Steel mills and aluminum smelters not only produced steel and aluminum; they also produced poisoned air. Even the marketable portion of their output sooner or later had to interfere with someone's pursuit of happiness. As discarded steel and aluminum objects accumulated, they either had to raise community trash disposal costs or contribute to environmental degradation, or both.

The extraction or growth of raw materials involves *people*. People are also involved in the manufacture of these materials into usable products, and in their distribution. Any change of resource use patterns therefore implies interference with people's accustomed activity patterns. In one generation (between 1940 and 1965), American production of manmade fibers per capita increased 420 percent. Whatever the advantages of these products to consumers, their increased production interfered with continuation of the accustomed life patterns of cotton and wool growers and of people employed in the cotton and wool processing industries. Per capita output of cotton in America declined 19 percent in this same generation, and per capita production of wool decreased 64 percent. Behind these figures lay hardship in the lives of thousands of people whose former niches were now oversaturated. Such people were compelled to undergo the trauma of re-adaptation to new niches, if they could find them.

Moreover, hidden under this chemical creation of better things for better living was another step in our commitment to reliance on ghost acreage and the drawdown method. Without recognition of its ramifications, an important change took place in the *kinds* of manmade fibers being produced. In 1940 only 1 percent of the manmade fibers produced were non-cellulosic. By 1965, the percentage had risen to 57. Cellulosic manmade fibers (rayon and acetate) used organic raw materials, so their production depended on supplies of renewable resources. The non-cellulosic manmade fiber production depended on supplies of *exhaustible* petroleum and coal. Thus we were unwittingly taking another step in the shift from a potentially interminable to a necessarily self-terminating way of life.

In comparison with an underdeveloped country, Americans were fortunate (so far) in having more freedom from hunger and more material wealth. But America's commitment to detritus consumption was farther advanced, and American life involved greater pressure in other respects. A good index of the extent of a people's commitment to a prosthetic and detritovorous way of life is their level of energy consumption. According to the United Nations *Statistical Yearbook*, by 1971 the per capita rate of energy consumption in the United States was about sixty times greater than in India. This made Americans more mobile than Indians (which is to say, more space consuming). We did more things. We made more varied demands on our environment and its material resources. We interacted more frequently and with more people in more ways. Americans had more ways of interfering with each other than most Indians did, though there was less competition for food between Americans than between Indians.

Each colossal American, with his phenomenally high rate of energy use, had, in effect, sixty times as many slaves working for him as did each Indian. So it would not have been too inaccurate to suggest that, for realistic comparison of American population *pressure* with that of India, the ratio between their respective population *densities* should have been multiplied by a factor of sixty. The number of actual people per square mile in the United States was about one-eighth of what it was in India. But if each American was about sixty times as colossal as each Indian, then American population pressure was equivalent to there being *twelve billion* people living the Indian pace of life on territory as large as the United States.

If there was consensus among Americans that "teeming India" was an overpopulated country, there ought to have been equally common recognition that the United States was too. Recognition lagged behind the fact because of persistent adherence to the old paradigm from the culture of exuberance.

Urbanization

The pressure was intensified still further for a large portion of the American and world population. Mass migration to the cities had been one expression of the revolution of rising expectations. Growth of the cities was due in various ways to growth of industrial technology.[7] But the congested and hypercompetitive life of the swollen cities contributed to the revolution of rising frustrations. In the affluent nations, many people attempted to have it both ways, earning their living in a central city but residing in an initially more spacious outer suburb— sometimes only to find the residential suburb swallowed up in an expanding metropolis.[8] The efforts of those who could afford commuting to escape from the high-density inner city clearly indicated that urbanization had produced unwelcome pressures. Use of fuel for commuting was also hastening drawdown.

Through urbanization, the people of a nation could be increasingly exposed to population pressure, even if their average density (over the nation's total expanse of territory) had not been growing. Since their average density *had* grown, then urbanization at the same time had to compound the pressurizing effect of that growth. For example, by 1960 53 percent of the people of the United States were living in 213 urbanized areas that comprised only $7/10$ of 1 percent of the nation's land.[9] Thus, while the mean density of U.S. population was close to the world average at about 50 persons per square mile, over half the nation was *experiencing* a mean density 75 times greater than that.

When settlement of the New World by Europeans was just beginning, the mean density of the Stone Age population of the territory that was to become the United States was about ⅓ person per square mile.[10] If the population had been evenly dispersed so that the average represented the actual experience of individuals, then each American Indian would have been within an hour's walking distance of about 17 other persons.[11] (Actually, since even the pre-Columbian population of America lived in groups, with unoccupied space between groups, the number of persons within an hour's walk of a given person—i.e., all the other people in his particular little settlement—would typically have been three or four times this number, occasionally even more. But the typical number of contact opportunities was clearly within the range where a person's life consisted mostly of dealing with persons one knew, not with strangers or anonymous functionaries.)

By the 1970 census, Alaska was the only state in which density remained so low. If evenly dispersed over the huge landscape, each Alaskan would have been within an hour's walk of 26 others. The least densely populated state among the "lower 48" was Wyoming, and in 1970 the hypothetical condition of even dispersal would have put an average resident within an hour's walk of 171 other persons. For the entire United States, the figure would have been 2,826—more than 166 times the pre-Columbian figure. For the most densely populated state, New Jersey, it would have been 47,909. But for the average resident of a central city in a U.S. metropolitan area, it would have been 402,128 (assuming that somehow, in spite of the obvious impediments to movement on foot that were implicit in urban density, such a person could still walk four miles in any direction in an hour). For the District of Columbia, where the nation's representatives were still trying to govern in the spirit of earlier times, the figure was 623,389.

Clearly, when people had so packed themselves into finite spaces, the opportunities for mutually impeding each other's movements and for otherwise interfering with each other's lives had become significant. Little wonder that the intimate kind of social ties that used to characterize human life in an era of agrarian villages had given way to an urban "norm of disengagement." Persons whom we physically encountered on streets, in subways, etc., evoked mainly the ability to stare past one another. Behaving toward each other as if persons were mere objects became a necessary defense mechanism because of the psychic overload imposed by urban circumstances.

By the 1970s so much of the world's population had been urbanized that the impact on the not-yet-urbanized portion was considerable. Before 1850 there had been no society that was predominantly urban. By 1900 Great Britain had emerged as the first such society. Less than one human lifespan later, most of the world was significantly influenced by urban ways, all industrialized nations were describable as "predominantly urbanized," and more than ⅕ of the world's people lived in nations that were more than 50 percent urban. This transformation was a profound feature of the end of exuberance. By the middle of the twentieth century, an unprecedented ⅙ of the world's population lived in urban agglomerations of 100,000 or more people apiece. In 1600, early in the Age of Exuberance, even Europe had less than 2 percent of its population living in agglomerations that size.[12]

Aggravated by urbanism, population pressure was intensifying competitive aspects of human interaction.

Pandemic Antagonism

Human efforts to "do something" about this predicament often made a bad situation worse. Agitation for change became common all over the world in this age of pressure. But frequently, in a crowded environment, the change sought by one group to solve a problem had to create new problems for others, who then demanded further changes to solve those problems. Unavoidably, those changes created still other frustrations for still other groups. In short, there was a general pattern of mutual interference. We needed to see that pattern, instead of being preoccupied with particular instances of conflict.[13] Pandemic antagonism was the ecologically expectable result of worldwide population pressure, voracious technology, and carrying capacity deficit.

One way we can acquire this global perspective is by simply divorcing the account of any particular episode from its locality or its specific identity. The reader could easily construct a "standard" news article for the age of overpopulation by taking an article about some symptomatic episode (of agitation or violence) from a given day's paper and blanking out only those words that identify the particular places, groups, or occasion. Such an article would be almost reusable another day if one were to fill in the blanks with different specifics. Stories of terrorist bombings, etc., become almost interchangeable, whether they emanate from Northern Ireland or southen Africa, from Asia or the Americas.

The specifics mask the important generalities.

Desired changes entail unwanted changes. Changed human activities involve changes in man's environment. Environmental change leads to succession; it can threaten human life. Non-competitive human interaction is imperiled by excess numbers and proliferating technology. Ecological antagonism begets social and emotional antagonism. These were the principles people needed to learn to read between the lines of the news in post-exuberant times.

In our exuberance, we proclaimed our yearning for peace on earth, and we professed good will toward men. But we multiplied and progressed—and found the world falling to pieces and men at each other's throats. As we became more and more numerous in proportion to any of the resources upon which we depend, we pressured ourselves into more competitive relationships and more antagonistic attitudes. Pressure has darkened our future. While we can hardly turn it off, we can somewhat mitigate its insidious impact by learning to understand how and why it has happened.

Notes

1. For illuminating comments on benefits accruing to New Zealand from its remote location, see Sinclair 1961.
2. Ehrlich and Ehrlich 1972, pp. 405–421.
3. Durkheim 1933 (listed among references for Ch. 6), p. 257.
4. Bureau of the Census, *Historical Statistics of the United States, Colonial Times to 1970,* Bicentennial Edition (Washington: U.S. Department of Commerce, 1975), I, 139–140.
5. Carl C. Taylor, *Rural Sociology: A Study of Rural Problems* (New York: Harper & Brothers, 1926), p. 21; italics added.
6. This and following comparisons based on *World Almanac* figures.
7. See, for example, Schnore 1965, especially Ch. 4; see also Davis 1955, 1965; Gibbs and Martin 1958.
8. Schnore 1965, Chs. 5, 6.
9. Davis 1965, p. 41.
10. Hauser 1969, p. 3.
11. This figure and those in the next paragraph are products obtained by simply multiplying the respective densities (persons per square mile) by the area of a circle with a four-mile radius, taken as an approximate hour's walking distance.
12. All figures in this paragraph are drawn from Davis 1965.
13. For a perceptive investigation of problems entailed by this pattern of mutual interference, see Milgram 1970. Cf. Galle et al. 1972.

Selected References

Davis, Kingsley
   1955. "The Origin and Growth of Urbanization in the World." *American Journal of Sociology* 60 (Mar.): 429–437.
   1965. "The Urbanization of the Human Population." *Scientific American* 213 (Sept.): 41–53.
Ehrlich, Paul R., and Anne H. Ehrlich
   1972. *Population, Resources, Environment: Issues in Human Ecology.* 2nd ed. San Francisco: W. H. Freeman.
Galle, Omer R., Walter R. Gove, and J. Miller McPherson
   1972. "Population Density and Pathology: What Are the Relations for Man?" *Science* 176 (Apr. 7): 23–30.
Gibbs, Jack P., and Walter T. Martin
   1958. "Urbanization and Natural Resources." *American Sociological Review* 23 (June): 266–277.
Hauser, Philip M.
   1969. "The Chaotic Society: Product of the Social Morphological Revolution." *American Sociological Review* 34 (Feb.): 1–19.
Milgram, Stanley
   1970. "The Experience of Living in Cities." *Science* 167 (Mar. 13): 1461–68.
Schnore, Leo F.
   1965. *The Urban Scene.* New York: Free Press.
Sinclair, Keith, ed.
   1961. *Distance Looks Our Way: The Effects of Remoteness on New Zealand.* Auckland: Paul's Book Arcade, for the University of Auckland.

VI | Living with the New Reality

The sociologist, no matter how gloomy his predictions, is inclined to end his discourse with recommendations for avoiding catastrophe. There are times, however, when his task becomes that of describing the situation as it appears without the consolation of a desirable alternative. There is no requirement in social science that the prognosis must always be favorable; there may be social ills for which there is no cure.
— Lewis M. Killian
The Impossible Revolution?, p. xv

The whole human enterprise is a machine without brakes, for there are no indications that the world's political leaders will deal with the realities until catastrophes occur. The rich countries are using resources with an extravagant disregard for the next generation; and the poor countries appear to be incapable of acting to curb the population increases that are erasing their hope for a better future. In such a world, declarations and manifestos which ignore the imperatives of the limits of growth are empty exercises. All the available evidence says we have already passed a point of no return, and tragic human convulsions are at hand.
— Stewart Udall, Charles Conconi, and David Osterhout
The Energy Balloon, p. 271

So long as the United States continues to assume that "more is better," all our efforts at increased energy efficiency—small cars, mass transit, industrial re-engineering—can achieve is to buy five, ten or perhaps fifteen years of additional time. . . . if we continue a self-indulgent, disposable society where the cycle continually is to dig, burn, build and then discard, we are stealing from our children and grandchildren the planet's resources.
— S. David Freeman
Energy: The New Era, pp. 330, 333–334

The basic features of a valid alternative technology have already been identified. . . . Unlike current bulldozer-supertanker technology, it would be based on ecological and thermodynamic premises that are compatible with the coexistence of man and nature over the long term. . . .
— William Ophuls
Ecology and the Politics of Scarcity, p. 126

13 | Backing into the Future

But We're Human!

We aren't *really* detritovores. A mere metaphor can't hurt us. After all, we're human. Crash can't happen to us.

How earnestly we would all like to believe that. But believing crash can't happen to us is one reason why it will. The principles of ecology apply to all living things. By supposing that our humanity exempts us, we delude ourselves. It is not just the yeast cells we put into wine vats that bloom. It is not just the recognized detritovores that crash. We have been backing into the future with our eyes too firmly averted from the detritovorous nature of our modern lifestyle. It is time to turn around and see what's ahead.

Whatever the species, irruptions that overshoot carrying capacity lead inexorably to die-offs. Irruptions can happen to any species that gains access to a previously inaccessible but highly suitable habitat.[1] All it takes is for the habitat to contain an abundance of whatever resources are needed by the invading species, and for there to be little population-checking pressure from predators and little or no competition from other species having similar niche requirements and living in the same area.

These final chapters provide no magic recipe for avoiding crash. There is none, when overshoot has already happened. It is in acknowledging that unwelcome fact that this book differs most fundamentally from previous ecological analyses. Facing that fact offers indispensable insights. Even writers deeply concerned with ecological aspects of the human predicament have remained strongly fettered by time-honored cornucopian thoughtways; many books have tried to persuade readers that if we will all become ecologically concerned in the nick of time, we may still avert the natural sequel to our excessive success.[2] We *didn't* become enlightened in time for that. But our post-exuberant predicament has had precedents, both human and non-human. We can and must learn from examining them, in order to face the future wisely.

It Has Happened to Humans

Bloom and crash happened to the people of Easter Island. Few of us in the northern hemisphere's industrial nations have ever had occasion to think about this triangular piece of land, 45 square miles in area, about as far out into the Pacific from the coast of Chile as Hawaii is from California. Nevertheless, its history has a vital lesson for us.

Easter Island is 1,400 miles from the nearest other humanly inhabited island. Around the time of Christ, one or two canoe loads of Polynesians happened to land there, where no human beings had previously lived. Among the things they had on board were some live chickens; so, although it seems doubtful that they could have previously known of the island's existence and it seems likely that they happened upon it fortuitously, they had evidently been prepared for a long journey—as if they might have been compelled to emigrate from some already overloaded island elsewhere in Polynesia. At any rate, they found here in the southeastern Pacific a fertile uninhabited island suitable for human takeover. They settled down to raise chickens, to garden, and to fish the adjacent waters, and they came to call their new home "Te Pito o te Henua," or "The Navel of the World."

An anthropologist, William Mulloy, extensively studied the island and its present inhabitants, descendants of these people.[3] He also helped reconstruct some of the great seaside altars the people once built, and to re-erect a number of the huge stone statues they carved during their time of highest cultural florescence. (The statues had been maliciously toppled during the iconoclastic episode of genocidal warfare that served to initiate the crash of this population.)

Some centuries ago, wrote Mulloy, when these people dominated their landscape so effectively, they

> must have felt great confidence in the future and a powerful sense of the impregnability of [their] accomplishments. As is twentieth century man, the Easter Islanders were technologically successful. Secure in the protection of the supernatural power of their deified ancestors who lined the shores in an unbreechable bulwark against the mysterious dangers of the empty seas and gazed pridefully inland upon the achievements of their issue, these industrious islanders must have rejoiced in the solid assurance that their success was permanent.

But disaster hovered and it was not precipitated by enemies from beyond the seas. Forty-five square miles was a finite environment and, with ever-increasing labor-consuming emphasis on religious construction, food had to be produced continually more efficiently by those allotted the task. Food-producing potential was probably never completely exhausted, though its limits may have been approached. It was, however, dependent on the uninterrupted maintenance of what must have been a highly coordinated social mechanism. Even slight disruption might have been expected to be sharply felt by many people. A legend [told by their living descendants] describes trouble and dissension erupting from disagreements about the idea of improving the productivity of agricultural land by removing surface stones and throwing them into the sea. Animosities once generated appear to have produced their usual reactions, and eventually two groups . . . fought a great battle along an entrenched line on the slopes of the volcano Poike. [One group was] said to have been all but exterminated.[4]

The resulting social chaos persisted, and the survivors continued to suffer high mortality from various causes.

Both by radiocarbon dating and by genealogical research, the time of this conflict has been set close to 1680. Survivors degenerated into continually warring bands who burned crops, molested fishermen, and actually hunted people for food. "To this day," said Mulloy, "the locality speaks eloquently of catastrophe, hopes unfulfilled, and projects suddenly abandoned. Hundreds of gigantic works of art remain unfinished and thousands of stone adzes and picks still rest where they were dropped by the artisans."

By the time of the island's European discovery (on Easter Sunday, 1722), its population had declined to an estimated 3,000 or 4,000, from a maximum that was probably at least twice that. Just before crash began, therefore, the population density had been roughly comparable to the 1970 average density for Michigan and Indiana. Mortality continued to exceed births, and the entire population of the island was down to 155 persons by 1886.

The subsequent demographic experience of these islanders gives little cause for hoping that *Homo sapiens* learns much from such experiences. By 1900 the population had increased to 213. By 1934 it was up to 456. By 1955 it had reached 842. In 1969 it was 1,432, and when Mulloy wrote (his account was published in 1974) there were 1,619 inhabitants of Easter Island. The average rate of increase during this period has been more than 3 percent per year, which means the population doubles in just about one generation.

For the rest of the world, in an age of *global* overshoot, the task facing mankind is to minimize the severity and inhumanity of the crash toward which we too are headed. Presumably mankind has found high death rates less unbearable when due to natural causes (e.g., microscopic predators) than when imposed by arrogant, sadistic human executioners. The function of ecological enlightenment at this late date is to enable us to do better than the genocidal factions on Easter Island. If, having overshot carrying capacity, we cannot avoid crash, perhaps with ecological understanding of its real causes we can remain human in circumstances that could otherwise tempt us to turn beastly. Clear knowledge may forestall misplaced resentment, thus enabling us to refrain from inflicting futile and unpardonable suffering upon each other.

Learning to Disregard Deceptive Differences

When irrupting populations surpass the newly available carrying capacity, the ensuing crash may occur by different means among humans, animals, or plants, although none are exempt from die-off. Access to new carrying capacity may come about differently for plants than for animals, and differently for other animals than for humans. But these differences do not affect the basic principle: die-off is the sequel to overshoot.

When die-offs occur among multi-celled plants, for example, the specific causal agents may be different from those that operate on yeast, or on algae. According to Elton,[5] a Canadian water weed, *Elodea canadensis*, was accidentally imported into Britain on American timber in the 1840s. At first it exploded into rivers, canals, ditches, lochs, and ponds all over the country, but then it declined after the 1860s "and has never again been considered a real plague. . . . The reasons for its decline . . . could be genetic, or indicate the exhaustion of some rare food element." Clearly, differences between algae and multi-celled weeds do not exempt either type of organism from crash as overshoot's sequel. Likewise we cannot count on being exempted just because we, too, are different.

Take another example, from the animal kingdom. In 1944, 29 reindeer were introduced to an island about 7,000 miles northwest of the scene of that human irruption-and-crash episode on Easter Island. The colonizer reindeer were placed on St. Matthew Island, an area of 128 square miles in the Bering Sea, quite suited to support them.[6] In 1957, 1,350 of them were counted. By 1963, the herd had increased to 6,000. Estimates of reindeer carrying capacity for land and climate similar to St. Matthew Island vary between 13 and 18 head per square mile, making the carrying capacity of this habitat between 1,600 and 2,300 reindeer. The 1963 population was thus at least 2.6 times what the island could permanently support. At least 3,700 of the living reindeer in 1963 were redundant. This did not mean, however, that as soon as that many had died off, the population would stabilize at the island's carrying capacity. *Overshoot leads to habitat damage,* so crash plummets population to a level *below* that which it might have sustained had it not overshot. An overgrazing herd steals from its own posterity. In 1966, only three years after the peak number was reached, there were just 42 reindeer left on St. Matthew.

Reindeer are as different from *Elodea canadensis* as the latter weeds are from algae and yeast cells (or as we are from reindeer). But when the conditions were right, they underwent the irruption–die-off sequence.

No Protection

We can see from these examples that the agent by which post-irruptive crash occurs may be something other than starvation. True, the Canadian water weed in Britain may have died back from exhausting some nutrient, and the reindeer devastated the lichen population upon which they depended for winter forage on St. Matthew Island. But the people of Easter Island *began their own crash by conflict,* and then inflicted starvation upon the survivors by maliciously disrupting food production activities. On the other hand, a study of an irrupting herd of Sika deer on James Island in Maryland found that they were well-nourished and parasite-free at the time of the die-off. Their growth was markedly inhibited, and this inhibition was shown by autopsy to be due to physiological disturbances induced by the *behavioral stress* associated with high population density.[7]

In drawing lessons for mankind at large from these varied examples, we must first acknowledge that species differences are no protection from the basic pattern. Second, we must recognize that, even with an abundance of food, crash can happen. Third, as we saw in Chapter 10, we must also recognize that the organized activities so indispensable to supporting huge populations by advanced-technology civilizations can and do break down. Conflict between factions within nations, as well as conflict between nations, can seriously reduce carrying capacity, as Sumner told us (see Chapter 5).

Queuing and Queue-Jumping

As Mulloy noted for Easter Island, so it is for modern nations: adequate sustenance production depends on highly coordinated social mechanisms. These can be fragile. If they break down, effective carrying capacity is diminished, leaving a swollen population in a state of overshoot.

To understand how human social processes were being revolutionized by ecological forces, we must examine a distinctively human form of social behavior—forming queues and, with mutual restraint, taking turns. Visualize a large city bank on a busy day. Customers sometimes arrive in the bank faster than the tellers can serve them, and queues spontaneously form.[8] Customers generally do a reasonable amount of waiting in line with reasonable patience, knowing others will similarly wait their turn to do business, and confident that the delay is no threat to completion of the business they have in mind.

Under post-exuberant conditions (outside of banks), that kind of faith was breaking down. To understand why, consider the quintessentially human quality of queuing behavior. It uses moral restraint to maximize the average fulfillment of self-interest, and it depends on cognitive comprehension of the situation and faith in reciprocity. The burly do not normally elbow their way ahead of the frail. The customer who happens to have personal acquaintance with a teller does not normally take advantage of it. A few exceptions to strict queuing may be tolerated in recognizably special circumstances, but if there come to be too many exceptions to the normal pattern of turn-taking, people will cease to infer that each exception is legitimate. Faith in the general fairness of the system gives way to doubt, and orderly queues may dissolve into a mad scramble.

The mad scramble could be only a matter of irritating inconvenience if everything else in the bank is functioning adequately. But if the reason customers were unwilling to wait in line arose from fear of the bank's becoming insolvent before they could take their turn at the counter, then the scramble could be far more serious. As a matter of fact, banks used to be *driven* into insolvency by panicking customers rushing to withdraw their money when anything made them fear that too many others might beat them to it.[9] Prior to the invention of deposit insurance schemes (which pool the reserves of many banks to protect each bank from collapsing when there is a run on its resources), bank panics were real and not uncommon.

In earlier chapters, nature's deposits of exhaustible resources were compared metaphorically to bank deposits. In the post-exuberant world, the bank of nature was being progressively subjected to a run or panic. The growing fear that waiting one's turn meant permanent deprivation because ultimately not all customers were going to be served was a major reason for the crescendo of demands for this, that, and the other thing "now!"

There is no system of deposit insurance for the bank of nature except biogeochemical recycling.[10] We have seen in previous chapters how enormously the rate of deposit was overshadowed by the modern rate of withdrawal. If man had confined himself to reliance on renewable resources, worldwide panic need not have begun to happen. But neither would we have achieved the high longevity and high affluence we have recently equated with the American dream. The fact that such panic did begin was due to man's enormous and still growing dependence on resource deposits that are exhaustible—on savings, as contrasted with income.

As we began to sense that we were living in an age of overpopulation, *all* human beings became threatened by the potential realization that *any* of us might be elbowed out of life's queue by our competitors. Regardless of whether we *said* the world was too full of people (or denied it), insofar as we *sensed* the intensification of competition, each of us was just as much in danger of being considered superfluous as the next person. The plight of the unwanted child became potentially everyone's plight.[11] There arose, therefore, an anxious quest for reassurance that somehow a status of redundancy applied only to others. This anxiety was at least part of what motivated militant social activism (sometimes combined with anti-social self-indulgence). While various crowds, demonstrating for or against various things, shouted various slogans, they were all implicitly saying, "It is you, not we, who are superfluous!" (Each particular crowd added some special epithet: ". . . you white racists!" ". . . you black bastards!" ". . . you squares!" ". . . you freaks!" ". . . you male chauvinists!" ". . . you fascist pigs!" The particulars, as usual, obstructed recognition of the general nature of the process.) Redundancy anxiety led to panic responses.

Panic is the opposite of orderly queuing. It happens when the confidence of society's members in each other breaks down and turn-taking gives way to an "every man for himself" type of scrambling behavior.[12] As sociologists had found, this can arise from (1) general perception of severe and immediate danger, coupled with (2) belief that the opportunities for escape are limited, (3) belief that these opportunities are diminishing, and (4) an absence of adequate communication about the danger.

219

Sociologists, no less than politicians and their once-exuberant constituents, at first stared without recognition at the accumulating symptoms of global panic in the post-exuberant age. Worldwide anti-social reactions were becoming common in response to premonitions of the impending crash that must follow our species' irruption.

"Freedom now!" "Stop the bombing now!" "Re-unite Ireland now!" "Halt pollution now!" "End apartheid now!" (As initials, the letters N.O.W. even designated one of the women's organizations that sprang up to liberate the female half of humanity—from constraints, pressures, and frustrations that were not as sex-specific in these post-exuberant times as the movement's ideology imagined.) In the 1960s and 1970s, many groups came to feel that their goals must be reached *immediately*. Some groups feared there might be no "eventually." With the future being more and more visibly stolen from, the reciprocity-maintaining virtues of compromise and forbearance appeared to turn into the vice of permanently lost opportunity. So demands became "non-negotiable." Queue-jumping became increasingly normal.

Elbowing and Counter-Elbowing

A generation earlier, when the Nazi debasement of humanity hit the world, we had not yet attained the ecological perspective that would have let us recognize the elements of queue-jumping in that ghastly episode. Now we must do so in order to understand *why* Germany's unconditional surrender in 1945 did not end the threat of man's de-humanization. Other great nations can march down the same tragic road.[13]

Imperial Germany had been a proud leader of industrial European civilization. Hindenburg, the final president of the ill-fated Weimar Republic, wrote in his last will and testament, dated May 11, 1934, that the German people had been required "to travel the road to Calvary." The world had failed to understand, he said, "that Germany must live . . . as the standard bearer of Western Civilization. . . ."[14] The crushing defeat in World War I and the humiliation that followed, aggravated by enormous economic hardships, had left millions of Germans deeply infected with a morbid need to reassert their significance. What happened thereafter in Germany was foreshadowed, we can now see, by the genocidal response to severe redundancy that had happened on Easter Island.

It was perhaps too early for an ecological interpretation of history that would have recognized *world* population redundancy as a source of social and political pathologies. Nevertheless, the victorious Allies had imposed upon their German foes World War I peace terms that said, in effect, "*Germans* are redundant." The burden of reparations payments was a heavy and continuing reminder of this message. An expectably defiant response was articulated in February, 1920, by the newly formed German Workers' party. One of its first seven members was Adolf Hitler; it became the Nazi party. Its program demanded not only "union of all Germans to form a Great Germany" with rights equal to those of other nations, and "abolition of the peace treaties of Versailles and Saint-Germain," but also "land and territory (colonies) for the nourishment of our people and for settling our excess population." In addition to this demand for access to ghost acreage, the nascent Nazi party demanded exclusion from German citizenship of anyone not "of German blood," particularly Jews, and the prevention of all non-German immigration into Germany.[15]

Nazism was thus, in part, a response to redundancy anxiety. That is why so many of the instances of political agitation and expressive activism in the 1960s and 1970s began to seem so remarkably Nazi-like to some crowd-watchers old enough to remember the 1930s.

The period 1933–45 showed to what depths of brutality human beings can sink when historic (or ecological) circumstances cause one group to declare another group superfluous. As in the Easter Island case, genocide resulted.

What Hitler and his henchmen told the despairing German people in the early 1930s was what various groups in other lands began saying to themselves in the 1960s with the clenched and upraised fist: "We are not redundant!" The Nazi version of the message went on to say the Jews *were* redundant, rationalizing this view and the resulting actions by asserting, in effect, "The Fatherland was unfairly elbowed out of its rightful position at the front of civilization's queue. We are therefore entitled to rebuild our strength and elbow our way back in." Substitute for "the Fatherland" any other resentful identity group—our race, our class, our profession, our sex, even our faith—and we are face to face with the general phenomenon of post-exuberant life for which we should have seen Nazism, an ominous special case, as a prelude.

After World War II we managed to convey the message that the superfluous ones were the Nazis, not Germans as such. Even the prolonged occupation and the division of Germany into two republics caught up in two separate big-power orbits hardly replicated the forces that unleashed Nazism. In Japan, too, defeat and occupation managed to say that the world could do without Japanese militarism rather than without Japanese people.

But if we did not replicate politically the forces that originally nurtured these perversions, they were nevertheless being created anew (in various places) by unacknowledged ecological processes. Queue-jumping and vicious retaliation against actions perceived as queue-jumping eventually became common features of postwar life. The resurgence of violent self-assertion the world over indicated that Hitler's "final solution" would not necessarily be the world's last experience with a genocidal effort by a mad tyrant or a frustrated people to elbow aside some scapegoat group on which they had decided to blame their misfortunes. Scapegoating was made likely by the pervasive frustrations of an age of colossal pressure. Elbowing and counter-elbowing by individuals, groups, and nations became commonplace.

These developments should have indicated that already the propensity of *Homo sapiens* for orderly societal living was being undermined by antagonism. As we know, human emotional antagonism may arise from ecological antagonism.[16]

The De-Civilizing of Homo sapiens

The longer we failed to understand this process of human self-destruction, the more we were fated to participate in it. For scores of centuries, unevenly and with many setbacks, human beings had been learning to pull in their elbows and be civil—i.e., to relate to each other not as colliding objects or competing animals but as compassionate humans. But the ongoing irruption of population and technology were compounding pressure until eventually this progress had to be undone. Competitive relations were to become again increasingly prevalent.

Population pressure made us react in pressure-increasing rather than in pressure-reducing ways, just as it had on Easter Island. As Sumner pointed out before the close of the nineteenth century, quarreling and discord would diminish the effective carrying capacity of the habitat. When population pressure imposed antagonistic relations upon us, therefore, our discordant responses squandered our efforts and our resources. We made our plight worse by our response to it.

For example, consider the following. At about the time of President Kennedy's "Ich bin ein Berliner" speech in West Berlin, there was in the United States a nationwide wave of construction of family fallout shelters. This was encouraged by the government as a means of enhancing the "credibility" of the nation's policy of nuclear deterrence. But it began to dawn on some families who had invested in shelters that, if their supposed prudence was less than universal, then "in the event of a thermonuclear exchange" grave problems would arise when families without shelters sought desperate entrance to their neighbors' shelters. Some people accordingly began to stock their limited-capacity shelters not only with canned food and water but also with shotguns to repel invading neighbors.

Ten years later this de-civilizing competition had escalated from an interpersonal to an international scale. In Australia, Sir Philip Baxter, the former chairman of that country's Atomic Energy Commission, predicted that nations in the northern hemisphere were almost certainly going to destroy themselves in a nuclear war. Survivors of such a war might invade Australia as a refuge from the uninhabitable north. "There will be far more than we can accommodate," he said, recognizing Australia's finite carrying capacity. "They will have no interest in allowing us to survive." In the same speech, he supported an Australian government decision to keep options open on the question of acquiring nuclear weapons. He gave this support on the grounds that such power could give Australians some say over who would be admitted and who would be kept out.[17]

It had become the kind of world, apparently, in which a continental shelter (of severely limited carrying capacity) might need a nuclear "shotgun" to repel desperate refugees.

This proposal came from no unreconstructed neo-Nazi in a twice-embittered Germany. It was the idea of an Australian—knighted heir to a tradition of British decorum, beneficiary of life in a young and vigorous nation. His country was large in area, large in self-esteem, in many ways more American than America in the 1970s.

Such developments showed how little objective basis there was for hoping we might counteract the erosion of hard-won empathy, the fragile foundation of civilization. Instead, as we became more frustrated in a world of increasingly pressuring shortages and intransigent competitors, we would probably become more and more exasperated. We were likely to be increasingly enmeshed in domestic instances of elbowing and counter-elbowing—political and ethnic conflict. And, difficult as it might be to believe it while we stood in the shadow of a war that so many Americans (and their friends overseas) had detested so vociferously, there was a real possibility that we would respond to frustrations from increased population pressure by reverting to war-like tastes. War may be hell—but under conditions of severe population pressure it can be attractive, because it offers relief from anxiety about our own redundancy. For the duration of a popular war, the enemy can be enthusiastically defined as the surplus people. If resources are squandered in the process, that makes the enemy all the more surplus.

In a habitat that was not growing any larger, the continuing increase in either our numbers, our activities, or our equipment would ultimately induce more and more antagonism. Our routine pursuit of legitimate aspirations as individual human beings, and as breathing, eating, drinking, traveling, working, playing and reproducing organisms, would increasingly entail mutual interference.

Notes

1. Assorted examples are described in Roots 1976 and Elton 1958.
2. As one example of a book that many readers might have dismissed as "alarmist" even though its avowed aim was to help "cure the growth disease," see Watt 1974. Similarly, according to Renshaw 1976, our task over the decades ahead was to "adjust" to the cessation of growth.
3. Mulloy 1974; see also a book translated from Spanish by Mulloy: Fr. Sebastian Englert, *Island at the Center of the World: New Light on Easter Island* (New York: Charles Scribner's Sons, 1970).
4. Mulloy 1974, p. 29.
5. Elton 1958, p. 115.
6. Klein 1968.
7. John J. Christian, Vagn Flyger, and David E. Davis, "Factors in the Mass Mortality of a Herd of Sika Deer *Cervus nippon*," *Chesapeake Science* 1 (June, 1960): 79–95.
8. For a mathematical analysis of varied forms of the queuing process, see Panico 1969. Early in the 1970s, many banks discovered that the necessary queuing was more effective and more equitable if all customers waited in *one* line, from which the front person could advance to the next available teller, instead of having a separate queue form spontaneously at each teller's window with no mechanism for assuring that all queues remained of equal length or moved at the same pace. Before this single queue system was installed, a well-run bank usually assigned some staff member to inform customers approaching a queue before a window that was getting ready to shut down and request them to join another line instead. Bank personnel would make that effort because it was understood that the workability of the system depended on customers' faith that when they lined up they would not wait in vain.
9. See Galbraith 1955 (listed among references for Ch. 10), pp. 184–185.
10. Cf. Commoner 1971.

11. For an elaboration of this plight, see Klapp 1969.
12. See Fritz and Williams 1957, p. 44; cf. Lang and Lang 1961, pp. 83–108; Turner 1964, pp. 409–413.
13. For concerned discussion of this possibility, see Phillips 1969, Ch. 5; cf. Ringer 1969 passim.
14. Snyder 1958, p. 422.
15. Ibid., pp. 393–394.
16. On conditions that nurture antagonistic interactions, see Russell and Russell 1968, and Sorokin 1942, 1975.
17. "Advice to Australia on Nuclear Weapons," *The Press* (Christchurch, New Zealand), May 19, 1972, p. 9.

Selected References

Commoner, Barry
1971. *The Closing Circle: Nature, Man and Technology.* New York: Alfred A. Knopf.
Elton, Charles S.
1958. *The Ecology of Invasions by Plants and Animals.* London: Methuen.
Fritz, Charles E., and Harry B. Williams
1957. "The Human Being in Disasters: A Research Perspective." *Annals of the American Academy of Political and Social Science* 309 (Jan.): 42–51.
Klapp, Orrin E.
1969. *Collective Search for Identity.* New York: Holt, Rinehart and Winston.
Klein, David R.
1968. "The Introduction, Increase, and Crash of Reindeer on St. Matthew Island." *Journal of Wildlife Management* 32 (Apr.): 350–367.
Lang, Kurt, and Gladys Engel Lang
1961. *Collective Dynamics.* New York: Thomas Y. Crowell.
Mulloy, William
1974. "Contemplate the Navel of the World." *Americas* 26 (Apr.): 25–33.
Panico, Joseph A.
1969. *Queuing Theory.* Englewood Cliffs, N.J.: Prentice-Hall.
Phillips, Peter
1969. *The Tragedy of Nazi Germany.* London: Routledge & Kegan Paul.
Renshaw, Edward F.
1976. *The End of Progress: Adjusting to a No-Growth Economy.* North Scituate, Mass.: Duxbury Press.
Ringer, Fritz
1969. *The Decline of the German Mandarins.* Cambridge: Harvard University Press.
Roots, Clive
1976. *Animal Invaders.* New York: Universe Books.

Russell, Claire, and W. M. S. Russell
   1968. *Violence, Monkeys and Man.* London: Macmillan.
Snyder, Louis L., ed.
   1958. *Documents of German History.* New Brunswick, N.J.: Rutgers University Press.
Sorokin, Pitirim A.
   1942. *Man and Society in Calamity.* New York: E. P. Dutton.
   1975. *Hunger as a Factor in Human Affairs.* Trans. Elena P. Sorokin, ed. T. Lynn Smith. Gainesville: University Presses of Florida.
Turner, Ralph H.
   1964. "Collective Behavior." Ch. 11 in R. E. L. Faris, ed., *Handbook of Modern Sociology.* Chicago: Rand-McNally.
Watt, Kenneth E. F.
   1974. *The Titanic Effect: Planning for the Unthinkable.* Stamford, Conn.: Sinauer Associates.

14 | Turning Around

Partial Reorientation

The third week in April, 1977, was a pivotal moment in history. It was the time when the world's most colossal energy users were at last called upon by their president to face the future realistically. In a Monday evening speech to the American people, in a Wednesday evening address to a joint session of Congress, and in a Friday morning nationally televised press conference, President Carter explained the need for, the philosophy behind, and the components of a comprehensive energy policy.[1] He sought to rechannel public thinking according to a new definition of the situation. In apparent recognition of the physical impossibility of continuing past patterns of escalation of energy use, emphasis in the proposed policy was on conservation.

Implicit in much of what the president said on those three occasions were some important indications that his mind and the mind of his chief energy advisor were in transition between paradigms. He was at pains to persuade his countrymen that the energy shortage was real, not just a contrivance of oil companies or other groups somehow able to profit from an illusory problem. The not-quite-explicit message of the week was the inescapable necessity for bold steps to prevent unnecessary aggravation of unwelcome realities.

By traditional American standards, the speeches were radical; i.e., they called for some rather basic changes in customary American practices. Some news media people understandably likened them to Winston Churchill's 1940 promise to the British people of a long, grim period of only blood, tears, toil, and sweat. In some respects, the comparison was apt. The world had waited too long by 1977 for a leader who would match the example of the man who became Britain's prime minister in the darkest days of World War II. Churchill had sensed his people's sober recognition of their nation's peril. He had forborne to belittle them with conventional political pep talk, and thus had summoned their indomitable courage by promising only what they already perceived was unavoidable.

Thirty-seven years later, as a spokesman for Realism, President Carter was taking a similar stance. He told Congress he could not give it "an inspirational speech" and did not expect much applause. But he himself had evidently become only partly reoriented to the new reality, and had only partly abandoned the old myth of limitlessness, for he adhered to long-standing tradition enough to tell the House and Senate that the "fair, well balanced, and effective plan" he was presenting could "lead to an even better life for the people of America."

Transitional Thinking

The president tried to make it clear that American problems were part of a global pattern. Early in the 1980s, he told Congress, "even foreign oil will become increasingly scarce." But he went on to note increasing overseas trade deficits, as if they were the essence of our predicament rather than mere ledger-book symptoms of the real problem: overshoot. "We imported more than $35 billion worth of oil last year, and we will spend much more than that this year. The time has come to draw the line."

Mr. Carter had not *wholly* grasped the new ecological paradigm. He let Congress persist in the old-style notion that trade deficits from increased oil imports were the source of national vulnerability. The vastly deeper danger that needed to be pointed out was that *further crowding of an already overloaded world would make us all more implacably competitive*. Competition for the world's remaining fuels was only a piece of that monumental problem.

Rampant competitiveness in all phases of human life (a war of each against all, as depicted long ago by Thomas Hobbes) was the horror we must now make concerted effort to try to prevent. The president's repeated emphasis on the need to be fair in the ways we apportion sacrifices showed that he sensed something like this Hobbesian threat. His thinking and planning had *begun* to be tuned in to the ecological paradigm.

It was bold for the president to point out that, even though the drawdown of world fuel reserves could be ignored for a while longer ("as many have done in the past"), ignoring it "would subject our people to an impending catastrophe." Only six years before, the American Petroleum Institute had begun a $4 million campaign of television and newspaper advertising, using the slogan, "A country that runs on oil can't afford to run short."[2] Now it was almost as if the federal administration's aim was to get Congress and the people to see the truth of precisely the opposite idea—a nation (or indeed a world) that is bound to run short of oil can't afford to run on oil.

However, the president's boldness was incomplete. A national leader as innovative as the time demanded, *fully* emancipated from the old paradigm inherited from the Age of Exuberance, would have warned his constituents of something more insidious than catastrophe: in the foreseeable future we shall feel more and more thwarted and subject to capricious manipulation by forces we cannot control, not necessarily because there will be more tyrants or unscrupulous profiteers, but principally because there will be more people, more resource-hungry technology, and more man-made substances crowding a world of fixed size.

The transitional, rather than altogether ecologically enlightened, nature of the president's thinking was apparent when he explained that "Along with conservation, our second major strategy is production and rational pricing." Although he was ecologically realistic in pointing out that "We can never increase our production of oil and natural gas by enough to meet our demand," his continued use of the customary word "production" to refer to *extraction* of fossil fuels reflected lingering influence by the obsolete paradigm. He called for a sensible system of fuel prices that would discourage waste and encourage "new production," not yet recognizing that the real meaning of "new production" was *hastened drawdown*—further reliance on the method that had brought us into our present grave predicament by providing seductive but temporary supplements to carrying capacity. His plan to shift from dependence on scarce oil and natural gas to "abundant" coal was, at best, a merely transitional remedy (and maybe he knew it). It left intact the exuberant expectation that savings withdrawals could be treated as income. It was simply a shift from using withdrawals extracted from a smaller savings depository to using withdrawals taken from a larger one.

Persistence of Obsolete Thoughtways

The boldness of President Carter's position was underscored by the way others clung more steadfastly to the old cornucopian assumptions. The minority leader in the Senate, Howard Baker of Tennessee, in an interview on ABC television immediately after the joint session commended the president's courage in presenting a comprehensive plan, and lauded some of its parts. But Senator Baker went on to say:

> The only thing that *really* bothers me about the president's proposal is it seems to me that we're *giving up*. We're giving up on the idea that we can produce ourselves out of this problem. And you know that's the traditional American way, that you *can* find new reserves of oil and gas, you can find new techniques to fuel the energy requirements of this country, and I'm not willing to give up. I'm not willing to allocate the shortage, and I'm not willing now to say that we've got to impose a huge tax on fuel oil or the equivalent of that tax on natural gas or gasoline in order to meet our requirements. There—in the oil and gas field, for instance—the last geological survey I saw by the government in 1975 indicated that there's as much oil probably yet undiscovered as we've ever found or used in the United States. And I find nothing in this proposal that creates any incentive for anyone to explore and develop those reserves.[3]

Senator Baker was obviously *not thinking by the new paradigm*. He did not see that efforts to provide incentives "to explore and develop" yet undiscovered deposits of fossil fuels would be efforts to continue the tragically prodigal hunting and gathering way of life in which the Industrial Revolution had so firmly entangled us. The Senate minority leader clearly had not yet accepted the idea that the world is finite, for he concluded by saying, "I don't think we have to settle for the bleak picture that the president painted."

From the same kind of myopia, the chairman of the Republican National Committee characterized the president's speech as "anti-consumer" and said,

> It is going to deter production. Its total emphasis is on conservation by raising taxes so as to inhibit consumer demand, and I think that's a very unfair and a very dangerous way to achieve an energy solution. In effect it says let's solve the problem by having no growth, by having inflation through government taxation, but without dealing with the basic problem of production, which is the traditional way of Americans to solve a problem, to draw the creative genius of the society to bear on the problem, to produce our way out of it, and to deal with it.[4]

Such appeals to "the traditional American way" reflected non-recognition that life was now being lived under post-exuberant circumstances. Traditions from an Age of Exuberance were obsolete. Leaders of the political opposition remained less able than the president to see the seriousness of the situation confronting humanity; they persisted in supposing options were still open that were not. They just did not seem to comprehend what it meant for mankind to have already overshot carrying capacity and to have already been so prodigal in drawing down finite resources. They meant to insist on more of the same. They still supposed we could permanently enjoy a carrying capacity surplus, that it was perpetually expandable. Thinking at least as Cargoists, and speaking almost like outright Ostriches, their "solution" would worsen our plight.

Questions for an Old World

Profound as it might seem by standards from the culture of exuberance, if the debate about how to cope with the future was going to resolve itself into merely an argument over how to "produce" our way out of trouble, the essential nature of our predicament would be overlooked. As it has been necessary to say repeatedly already, overlooking that predicament could not protect us from it. What really needed to be discussed was not only the dire need to conserve resources, but also this: *What kind of role are human beings going to play in their own impending crash?* How much will our efforts to avoid the unavoidable make it worse?

Men who continue to perceive our predicament according to a pre-ecological paradigm simply will not recognize limits imposed by our world's finiteness. They are thus required by their assumptions to suppose that only the machinations of antagonists can thwart our attainment of long-sought goals. People who think that way can be tempted to take up arms against a sea of troubles—troubles which weapons have no power to end. In so doing they will only add to the future's tempestuous onslaught.

If instead, guided by knowledge based on the new paradigm, we can face reality, we may recognize that we still could make some adjustments to stem the tide of our de-civilization. Those adjustments will *not* "lead to an even better life," but they may keep us from making our future more gruesome than it has to be. To see what really needs to be done, we must ask ourselves several excruciatingly tough questions. They carry our thinking far beyond the point reached in political discussions of energy policy.

(1) First and most fundamentally, we must ask whether we can begin making ourselves less detritovorous. *Can we begin to phase out our use of "fossil fuels" as combustible sources of energy?* Can we actually begin to wind down our use of oil, natural gas, *and* coal as fuels?

It would be hard to state an imperative that goes more diametrically against the grain than this. Far more drastic change is required of us by post-exuberant circumstances than merely shifting from scarce oil to "abundant" coal, and supplementing both with nuclear power, as proposed by the president.[5]

Our initial response to the energy pinch during the harsh winters of the late 1970s was to increase efforts to extract the earth's Carboniferous legacy. In 1977, not only a culprit-seeking Congress and the scandal-hungry news media but even the conservationist Secretary of the Interior, Cecil Andrus, felt obliged to berate oil and gas companies for "withholding" some of nature's finite legacy rather than "producing" it as rapidly as possible. During the coal miners' strike in 1978, midwesterners became alarmed as stockpiles of already mined coal above ground were drawn down, but few learned to fear national dependence upon ongoing drawdown of unmined underground stocks. Habituated to unrecognized prodigality, Americans breathed a sigh of relief when the miners went back to work "replenishing supplies"— i.e., resuming extraction of nature's exhaustible legacy.

What was needed was national redefinition of such substances, involving abandonment of our deep-seated assumption that their use *as fuels* is inherent in their nature. It is high time to learn (before we become even more numerous and still more "developed") that the wisest "use" of coal and oil may be to leave them underground as nature's safe disposal of a primeval atmospheric "pollutant"—carbon. By our ravenous use of fossil acreage to extend carrying capacity we not only prolonged human irruption but also began undoing what evolution had done in getting the atmosphere ready for animals (including man) to breathe, and ready to sustain the kind of climate in which present species (including ourselves) had been evolved. Hundreds of millions of years of evolution had produced the oxygen-rich and nearly carbon-free atmosphere we need, and had apportioned the earth's supply of H_2O between atmosphere, ice caps, and oceans.

Now mankind seemed bent on undoing in just a few centuries what nature had so slowly accomplished. Human actions have appreciably changed the CO_2 and dust content of the atmosphere. This may appreciably alter world climate. Change in either direction away from the recent optimum may seriously reduce world agricultural output. Added CO_2 can trap heat and may melt the polar ice caps and raise sea level, with disastrous effects on human settlement patterns. Conversely, added dust, for all we know, might screen out enough sunlight to hasten a new Ice Age; depleted fuel reserves would make mass adaptation to that situation quite impossible.[6] We need to accept the earth as it was when our species evolved upon it. Had it been different, *Homo sapiens* could not have emerged.

(2) The next most fundamental challenge to consider is whether enough of us can recognize at last the inescapable intricacy of any non-detritovorous relationship between the human species and its habitat. We urgently need to realize, as did the forester Aldo Leopold (who grasped the new paradigm as long ago as 1933), that if a civilization is to endure, it must be a system of "mutual and interdependent cooperation between human animals, other animals, plants, and soils." We need to share Leopold's recognition that despoliation of the natural environment is no mere annoyance but "has evicted nations, and can on occasion do it again. As long as six virgin continents awaited the plow, this was perhaps no tragic matter—eviction from one piece of soil could be recouped by despoiling another. But there are now wars and rumors of wars" (he said in the year Hitler came to power) "which foretell the impending saturation of the earth's best soils and climates. It thus becomes a matter of some importance, at least to ourselves, that our dominion, once gained, be self-perpetuating rather than self-destructive."[7]

Barring human extinction, there will never come an end to man's need for enlightened self-restraint—the conservation ethic, as Leopold understood it. Yet the abandonment of self-restraint has been naively advocated during frustrating times by various self-appointed world-savers of either Cargoist or antinomian persuasion. Their destructive counsel must be flatly rejected.

(3) If we come anywhere near measuring up to these first two challenges, we must then ask whether we can candidly acknowledge that general affluence simply cannot last in the face of a carrying capacity deficit. That fact is perhaps only a trifle less repugnant than the idea that the buried remains of the Carboniferous period must not be taken as fuels. There is much resistance to it. Amid congressional debate of the administration-proposed national energy policy there was public yearning to believe that curtailed use of energy need not entail reduction of the traditionally high American standard of living. In an admonitory (and perhaps wistful) way, it was noted that affluence of an American level somehow prevailed in Sweden and in West Germany despite their using only about half as much energy per capita.[8] As recognition grew that perhaps *we* could get by with smaller cars, etc., it was hoped that increased efficiency in all aspects of modern technology might suffice to dissolve the problems arising from our civilization's commitment to the drawdown method.

In this spirit, some scientists had already suggested reforms in energy use very similar to the president's objectives. Two physicists had proposed in late 1976 a very substantial list of changes which, all told, could reduce American energy consumption by about 40 percent.[9] Their list called for such heroic efficiency steps, however, that it seemed obvious neither public nor industry would willingly undertake or fully implement that much reform:

a) all residential heating by electrical resistance devices to be replaced by heat pumps having a coefficient of performance of 2.5 (i.e., able to "pump" two and a half times as much energy as it takes to run them);

b) home air conditioning equipment to be increased to a coefficient of performance of 3.6;

c) home refrigerators to be made 30 percent more efficient;

d) household water heating fuel requirements to be reduced by half;

e) home insulation to be improved enough to reduce heat losses by 50 percent;

f) home air conditioning loads to be reduced by reducing infiltration;

g) "total energy systems" to be installed in half of all multi-family dwellings and in one-third of all commercial buildings;

h) microwave ovens to be used for half of all home and commercial cooking;

i) air conditioning and refrigeration in the commercial sector both to be made 30 percent more efficient;

j) use of fuel for commercial water heating, and for lighting and heating commercial buildings to be cut in half;

k) insulation of commercial buildings to be improved enough to reduce air conditioning demand by 10 percent, and ventilation rate to be cut in half and heat recovery used to reduce air conditioning demand another 15 percent;

l) better management practices to be applied in the industrial sector;

m) other major improvements to be made in efficient use of fuel, electric power, heat, and steam, by industry;

n) aluminum, iron, and steel components of urban refuse to be recycled;

o) organic wastes in urban refuse to be used for fuel;

p) throughput at petroleum refineries to be reduced, and field and transport losses in use of natural gas to be reduced;

q) automobile fuel economy to be improved 150 percent;

r) fuel consumption by other transportation systems to be reduced 35 percent.

Monumental as this list of efficiency measures might seem to a country reluctant to curtail affluent habits, the two physicists acknowledged that *even 40 percent reduction of energy use would merely buy us 17 years' postponement of the ultimate consequences of our colossal commitment to drawdown.* If we do make ourselves stretch out the drawing down of exhaustible resources, it will undeniably delay the darkening of our future and it will slow the rate at which we steal from our descendants. It is worth doing, but there is wishful thinking involved when we suppose that the pursuit of energy efficiency by innovations in engineering will also enable us to "eat our cake and have it too."[10] That supposition needs to be recognized as mainly a last-ditch version of Cargoism.

(4) Depletion of ghost acreage is not only forcing us to take stringent efficiency measures, but it will also irresistibly compel return to a simpler life. Will we accept it with any grace? Or will we kick and scream our way into it, imagining we could always have everything we want if only those government people weren't forbidding it? Very few of the counterculture adherents who tried during the Vietnam war years to abandon a materialistic culture they had come to abhor were able to create any lastingly viable alternative. So it seems hardly probable that mankind on any large scale will adapt gracefully to "de-development." The urge toward worldwide development remains strong and pervasive. We are long out of tune with the mores of asceticism.

We do need somehow to nurture the faith that a reduced "standard of living" might bring, in exchange, for some people, some positive gains in the "quality of life." But it will be a difficult faith, much easier to preach than to practice.

(5) Is there any chance that we can learn to practice such mandatory austerity unless we can first be spared the widespread, deliberate badgering of people into wanting more, more, more? With the new paradigm we should begin to recognize the increasingly anti-social ramifications of advertising. We need to discredit and wind down this want-multiplying industry, perhaps even legally suppress it. In an overpopulated and resource-depleted world, an industry fundamentally devoted to making people dissatisfied with what they have, however respectable such an enterprise has come to seem by standards derived from pre-ecological thoughtways, is an industry dedicated to augmenting human frustration. In an age of overshoot it is bound to foster the resentful attitudes that could turn inescapable competition into destructive conflict.

Advertising did not always loom so large in our lives. When there was still an American frontier, there was virtually no advertising industry at all. It should not be inconceivable that we could do without such badgering. Freedom to exhort people to hasten exhaustive drawdown of nonrenewable resources was not what the authors of the American Bill of Rights had intended to establish. It is, moreover, an inherently self-terminating kind of liberty, for it fosters the self-destruction of our hunting-and-gathering industrial economy. We need to free ourselves from its compulsions.

Make no mistake, though; this would entail truly revolutionary changes. It calls for drastic reorganization of American television and radio broadcasting and newspaper and magazine publishing. It would be vehemently resisted. In other nations, where these media are not so exclusively financed by advertising revenues, the readjustment might be marginally less drastic. But in the United States it would require an appropriately post-exuberant reinterpretation (if not outright alteration) of the First Amendment provision for unabridged freedom of speech and press. It is time to recognize, however, that these freedoms were not meant to authorize irresponsible encouragement of habitat destruction. *Homo sapiens* was far less prosthetic, far less detritovorous, and far fewer in numbers when the First Amendment was added to the U.S. Constitution.

Sapiens? Radical?

Have we the wisdom (as our species name implies) to think through questions this radical?

The new requirements of adaptation to post-exuberant circumstances far exceed what the old non-ecological orientation enabled our minds to accept or even to contemplate. Even those who profess to be "radical" or "revolutionary" (in the merely political sense) seldom penetrate to the real root of the human predicament. Working yet from an obsolete paradigm, they never speak of overshoot. The myth of limitlessness still pervades their rhetoric.

Self-styled revolutionaries imagine themselves to be very radical when they propose merely to get rid of capitalist procedures for providing burgeoning populations with economic progress by accelerated drawdown, and to replace them with socialist procedures for providing burgeoning populations with economic progress by accelerated drawdown. In an ecological sense, they leave the problem intact. Proposals are not radical if they accept the continued drawing down of non-renewable resources; there is nothing radical about proposing to speed up that crash-inviting process.[11]

A *truly* radical mind would have seen that the public debate launched by President Carter's energy policy messages was ultimately a debate over whether to curtail human dominance over the global biotic community. That dominance had become dangerously excessive. Nevertheless, the debate languished. The first round took a year and a half to culminate in passage of energy legislation much weaker than the measures requested by the Carter administration three months after taking office.[12] Meanwhile, continued increases in fuel prices evoked complaints and apprehensions but gave little impetus to fuel conservation. The hard questions (posed in this chapter) were never faced. Thus it was far from certain that the outcome of any further rounds in the debate would halt the self-destructive activities of *Homo colossus*. Congress and the people were unaccustomed to seeing things ecologically and simply did not come to grips with the real issues affecting our destiny. The generally Cargoist tone of most discussions[13] actually diverted attention from life-and-death questions.

What was at issue was whether, after two hundred years of increasingly colossal reliance on drawdown, mankind was going to find it *possible* to revert to living on income. There might be already too many of us for this to be possible. Could four billion of us get back to depending on sustainable yields of contemporary renewable resources by putting aside our ingrained aspirations to become even more colossal? Would we actually have to become *less* colossal than we already were? How much so? How were we to decide who would give up what? By what standards of equity could such sacrifices gain acceptance?

If even de-development wouldn't return four billion of us to living on income, how, when, and with what severity would die-off befall humanity? Who would die early and who would live out a "normal" lifespan? If crash of one form or another had become unavoidable, were there alternative ways for it to happen, and was there reason to prefer one way over others? Could we take any steps to make the least inhumane way most likely, and guard against drifting into more horrible patterns?

Political and popular attention simply were not focusing on these questions. As we near the end of this book it should be clear *why* they were not. Bearing in mind the contrast between cornucopian and ecological paradigms, it is easy to understand why most public concern over what to do about "the energy crisis" either continued to proceed from doubt that the problem was real or persisted in debating the comparative merits of various ways of obtaining more energy—or maintaining access to customary supplies. Steps that would, in the final analysis, aggravate our predicament were still mistaken for solutions.

Persistence of an obsolete paradigm made national leaders and millions of their constituents respond like a generation of Rip Van Winkles, as if they had slept through a revolution. They did not seem to know how fundamentally different it was to be living in post-exuberant times, after internalizing expectations formed in an Age of Exuberance. The time had come for everyone to begin thinking about whether mankind has already overshot the earth's ultimate human carrying capacity. In this book I have tried to make it clear why this is a matter we dare not avoid facing.

Paradigm versus Paradigm

The world, man's place in it, and a host of policy issues all look different when viewed in ecological terms than when viewed in traditional, pre-ecological perspectives.[14] Let us therefore now try to encompass the new ecological paradigm in a few sentences, and then explicitly contrast it with the dangerously obsolete worldview it must replace. The ecological paradigm clearly recognizes the following basic ideas:

E1. Human beings are just one species among many species that are interdependently involved in biotic communities.

E2. Human social life is shaped by intricate linkages of cause and effect (and feedback) in the web of nature, and because of these, purposive human actions have many unintended consequences.

E3. The world we live in is finite, so there are potent physical and biological limits constraining economic growth, social progress, and other aspects of human living.

E4. However much the inventiveness of *Homo sapiens* or the power of *Homo colossus* may seem for a while to transcend carrying capacity limits, nature has the last word.

These precepts stand in sharp contradiction to some assumptions deeply imbedded in Western thoughtways by 400 years of exuberance. That is why the ecological paradigm has so little political acceptance, and why it has had so little influence upon public policy. The 400-year boom made the following assumptions plausible and popular:

P1. People are masters of their own destiny; they are essentially different from all other creatures, over which they have dominion.

P2. People can learn to do anything.

P3. People can always change when they have to.

P4. People can always improve things; the history of mankind is a history of progress; for every problem there is a solution, and progress need never cease.

These assumptions from the Age of Exuberance came to seem so self-evident that they even became incorporated (in ostensibly more sophisticated form) into a quite unecological paradigm that has guided the inquiries of most social scientists. For scholars in the social science disciplines, premises from the Age of Exuberance took on exaggerated importance because they played a part in the struggle of these new disciplines for academic recognition. Insofar as the supposed autonomy of man from nature seemed to provide justification for independence of social science from biology, the struggle tended to give social scientists an anti-biological bias. Only recently, therefore, have any significant number of us in the social sciences begun to see how the accumulation of solid biological knowledge calls for questioning cornucopian expectations.

The social science version of the four popular notions (P1 to P4) might be stated as follows:

SS1. Since humans have a cultural kind of heritage in addition to and distinct from their genetic inheritance, they are quite unlike the earth's other creatures.

SS2. Culture can vary almost infinitely and can change much more readily than biological traits.

SS3. Thus, since many human characteristics are socially induced rather than inborn, they can be socially altered, and inconvenient differences can be eliminated.

SS4. Thus, also, cultural accumulation means that technological and social progress can continue without limit, making all social problems ultimately soluble.

There are elements of truth in the popular assumptions, and the social science version of them may indeed have sharpened their valid points. But these elements (even in the social science version) are still susceptible of misleading exaggeration. To cope with post-exuberant circumstances we must be more critical of past premises. (1) We must now see that people are indeed different from other creatures, but *not altogether* different.[15] Our cultural type of inheritance is tremendously significant; it evolves in response to differently operating selection pressures than those that change genotypic distributions. It was, however, a gross exaggeration to suppose that culture exempted us from principles of ecology. (2) We must now also see that although people can learn, learning isn't always easy, and we sometimes learn inappropriate things. Our cultures do change, but not instantly, and not always in directions that are adaptive to existing or future conditions. Habits are hard to break; cultural inertia is hard to overcome.[16] (3) We must recognize that if people's expectations, actions, and even their sex roles and other supposedly inborn traits do change, sometimes this happens only under extreme pressure. And sometimes even in the face of necessity people do not change, or they change in futile ways. Moreover, even when people (as individuals) may have changed, or may be willing to change, whether their institutions will change may be quite a separate matter. Social organization can resist modification just as truly as can whatever individual traits happen to have a large genetic component.[17] (4) We need to acknowledge that although people have improved things, not every change people make is an improvement, and not all improving changes are free from unwanted side effects. So-called side effects can sometimes outweigh the advantages achieved by an "improvement." Egypt's Aswan Dam, promoting disease as well as achieving flood control,[18] may be a major case in point; we should also remember that the prescription drug Thalidomide was considered a blessed antidote to discomforts in pregnancy until its power to deform fetuses was belatedly discovered. Visible progress, social or technological, has often been accompanied by initially unrecognized harm. Past records of progress simply do not signify either the imminence or the immanence of the millennium. Uncritical extrapolation of trends without careful inquiry into the mechanisms that produced them can yield dangerously misleading forecasts.

In short, under post-exuberant circumstances we tend to be ill served by reasoning from the popular premises P1, P2, P3, and P4. We may not be much better served by their social science counterparts, SS1, SS2, SS3, and SS4. Instead, we need to see our circumstances in terms of propositions E1, E2, E3, and E4. And we need to regard them not as conclusions, but as starting points for a whole new line of thought about the human predicament.

If most of the words and phrases listed in the glossary of this book become incorporated into the daily working vocabulary of the reader, the ecological way of seeing will begin to take effect. I must emphasize, however, that this does not mean the reader will begin to see obvious *solutions* to the human predicament. To expect that adoption of an ecological paradigm will have this result is to think in Cargoist fashion. Cargoist expectations are an inappropriate criterion for evaluating the ecological paradigm this book advocates, just as they were an inappropriate yardstick for assessing the Carter administration's energy plan.

Notes

1. See the transcripts of these three presidential efforts in the *New York Times*: Apr. 19, 1977, p. 24; Apr. 21, p. B8; Apr. 23, p.6.
2. Freeman 1974, p. 192. I recall with a bit of nostalgia that at least one major oil company also printed the slogan on the road maps that its service stations were then still giving away free of charge.
3. Transcribed from a tape recording made at the time of the telecast. (Italicized words reflect emphases in Senator Baker's elocution.)
4. Transcribed from tape. (See note 3). For other indications that pre-ecological thoughtways have had an even firmer grip on the minds of Republicans than on the minds of Democrats, see Dunlap and Allen 1976.
5. Cf. Executive Office of the President 1977.
6. C. C. A. Wallen, "Man and Climate," *World Health,* Nov. 1973, pp. 12–17; Bach et al. 1979; Ponte 1976; Schneider and Mesirow 1976.
7. Aldo Leopold, "The Conservation Ethic," *Journal of Forestry* 31 (Oct., 1933): 634–643. Cf. Leopold 1949.
8. For example, see Schipper and Lichtenberg 1976; Mazur and Rosa 1974.
9. See Ross and Williams 1976, 1977. Also cf. Allen L. Hammond, "Conservation of Energy: The Potential for More Efficient Use," *Science* 178 (Dec. 8, 1972): 1079–81; Dumas 1976; Stobaugh and Yergin 1979, pp. 136–182.
10. Dumas 1976, pp. 263–264.
11. When Marxists diagnose the world's ills by concepts derived from a pre-ecological paradigm, their prescriptions remain more Cargoist than radical. They remain as blind as "the establishment" in capitalist nations to the perils of detritus dependency and to the principle stated in Ch. 1, that human society is part of a global biotic community in which excessive human dominance is self-destructive.
12. Even to note that it took Congress a year and a half to legislate in response to Carter administration requests is to understate congressional procrastination. As long ago as February, 1939, Congress had received from a previous president a report suggesting legislation required for prudent use and conservation of energy resources. In his letter of transmittal, President Roosevelt had pointed out: "Our energy resources are not inexhaustible, yet we are permitting waste in their use and production. In some instances, to achieve apparent economies today, future generations will be forced to carry the burden of unnecessarily high costs and to sub-

stitute inferior fuels. . . ." See "Energy: How Long Will It Take?" *Christian Science Monitor,* Feb. 8, 1977, p. 31.

13. President Carter himself seemed to turn more Cargoist in the second round of the debate, when he called for an all-out program to develop "synthetic fuels." See "The Energy Plan," *Newsweek,* July 23, 1979, pp. 27–28; "Carter's Energy Plan," *Newsweek,* July 30, pp. 54–58.

14. See William R. Catton, Jr., and Riley E. Dunlap, "Environmental Sociology: A New Paradigm," *American Sociologist* 13 (Feb., 1978): 41–49; Riley E. Dunlap and William R. Catton, Jr., "Environmental Sociology: A Framework for Analysis," Ch. 3 in Timothy O'Riordan and Ralph C. D'Arge, eds., *Progress in Resource Management and Environmental Planning* (Chichester, England: John Wiley & Sons, 1979), I, 57–85.

15. Literature has begun to accumulate that reflects renewed attention both to the sociological relevance of man's biological nature (e.g., Allan Mazur and Leon S. Robertson, *Biology and Social Behavior* [New York: Free Press, 1972]) and to the biological relevance of social patterns (e.g., David P. Barash, *Sociobiology and Behavior* [New York: Elsevier, 1977]).

16. See Ch. 10, notes 7 and 8.

17. See Richard T. LaPiere, *Social Change* (New York: McGraw-Hill, 1965), pp. 327–332.

18. Ehrlich and Ehrlich 1972 (listed among references for Ch. 12), p. 183.

Selected References

Bach, Wilfrid, Jürgen Pankrath, and William Kellogg, eds.
1979. *Man's Impact on Climate.* New York: Elsevier.
Dumas, Lloyd J.
1976. *The Conservation Response: Strategies for the Design and Operation of Energy-Using Systems.* Lexington, Mass.: D. C. Heath/ Lexington Books.
Dunlap, Riley E., and Michael Patrick Allen
1976. "Partisan Differences on Environmental Issues: A Congressional Roll-Call Analysis." *Western Political Quarterly* 29 (Sept.): 384–397.
Executive Office of the President, Energy Policy and Planning
1977. *The National Energy Plan.* Cambridge: Ballinger Publishing.
Freeman, S. David
1974. *Energy: The New Era.* New York: Walker.
Leopold, Aldo
1949. *A Sand County Almanac.* New York: Oxford University Press.
Mazur, Allan, and Eugene Rosa
1974. "Energy and Life Style." *Science* 186 (Nov. 15): 607–610.
Ponte, Lowell
1976. *The Cooling.* Englewood Cliffs, N.J.: Prentice-Hall.
Ross, Marc H., and Robert H. Williams
1976. "Energy Efficiency: Our Most Underrated Energy Resource." *Bulle-*

tin of the Atomic Scientists 32 (Nov.): 30–38.

1977. "The Potential for Fuel Conservation." *Technology Review* 79 (Feb.): 49–57.

Schipper, Lee, and Allen J. Lichtenberg

1976. "Efficient Energy Use and Well-Being: The Swedish Example." *Science* 194 (Dec. 3): 1001–13.

Schneider, Stephen H., with Lynne E. Mesirow

1976. *The Genesis Strategy: Climate and Global Survival.* New York: Plenum.

Stobaugh, Robert, and Daniel Yergin, eds.

1979. *Energy Future: Report of the Energy Project at the Harvard Business School.* New York: Random House.

15 | Facing the Future Wisely

Perilously Persistent Cargoism

It is essential to see the profound peril in continued flagrant misperception of the very nature of the human situation. That peril compelled me to write this book. Misperception is the problem to be overcome by a paradigm shift, and only a paradigm shift can overcome it. Misperception will tend to motivate efforts to pursue ostensible "solutions" that will, when the circumstances are so much different from what people suppose them to be, make matters worse instead of better. In this concluding chapter, therefore, I want to highlight the contrast between prevalent presuppositions about our situation (and about our options) and the actual ecological circumstances that need to be recognized and with which we must do our best to cope.

During the debate on energy policy, one ominous sign of the prevalence of mistaken assumptions was the fact that Cargoist expectations could pass for Realism. As we saw in Chapter 4, Realism and Cargoism are importantly different. This difference has been obscured during public debate by exaggeration of a lesser difference—between Cargoist expectation that "tech fixes" will overcome our problems, and the more exuberant former expectation that problems would naturally vanish with passage of time.[1]

In referring to Cargoist thoughtways I am not now merely speaking of particular technological proposals—of domestic solar water heaters or other "soft energy paths," of the breeder reactor, stepped-up oil exploration, better carburetors or smaller cars, gasification of coal, geothermal power generation, etc. At this point I have in mind *the general background belief* that carrying capacity can "always" be raised anew by further technological breakthroughs. As we saw in Chapter 2, major technological achievements did in past ages repeatedly raise the ceiling for human population. Modern Cargoism naively supposes this picture of the past must also be a valid picture of the future. It may not be.

FIGURE 1. Three Images of Growth

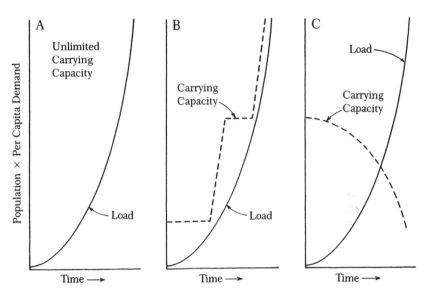

In Figure 1, three images are compared. Image A schematically depicts a curve of exponential growth in a situation where there is no ceiling. That image represents the expectations nurtured by the Age of Exuberance. Only the outright Ostrich can hold that view of the human situation today. Image B (if taken as a picture of the past) shows that Image A was once plausible because the ceiling sometimes was raised before the growth curve had quite bumped against it. The Cargoist way of thinking takes Image B as a picture of the future. Image C, a schematic way of reflecting the fact that prosthetic man's resource appetite has grown larger and larger, shows that carrying capacity can be considered a diminishing quantity rather than an ever-enlargeable quantity. Technological advances can *reduce* carrying capacity, even if historically their function was to raise it. Image C is much more different from Image B than the latter is from Image A. Insofar as Image C is applicable to mankind's situation in the real world of post-exuberant times, then Image B is *not sufficiently* more realistic than Image A to serve in charting our way into a post-exuberant future.

Despite this, there are persistent temptations toward Cargoist thinking. By the time the thirty-ninth president of the United States implored his countrymen to devote themselves to "the moral equivalent of war" to meet the challenge of living in a post-exuberant world, to tax themselves in ways that might begin to foster energy conservation, and by the time he urged Congress to enact various efficiency-promoting policies, other developments were already beginning to provide excuses for supposing crash would be averted. Hope continued to spring from anticipation of new ways to accelerate drawdown.

Oil companies, styling themselves "energy companies" as they diversified their holdings and their entrepreneurial activities, filled the media with ads extolling technological breakthroughs to come. Americans would beat the Arabs at their own game; if the Arabs happened to be sitting on top of the world's largest remaining oil reserves, Americans had more than that much energy under their own land in the form of coal. The ad writers glowingly told the viewing and reading public about new ways being devised to dig it up faster and burn it better. Coal would be liquefied, gasified, piped in one form or another to where we wanted it. (The need for vast quantities of water to do this, and the scarcity of water in some areas where much of the strip mining was going to occur, were not mentioned in the ads. Nor was the inescapably dead-end nature of the drawdown method ever acknowledged, amid Cargoist rhetoric and imagery.)

Excuses were available internationally for thinking along Cargoist lines. European nations had faced a serious emergency at the time of the 1973 Arab oil embargo, but this had subsided. Now, oil rigs in the North Sea were "producing." New life was being pumped into the British economy by the oil being pumped ashore in Scotland. Norway, too, was extracting North Sea petroleum, and was rising in affluence among the top-ranking nations. So neo-exuberant attitudes were again temptingly plausible.

In America, shortages of gasoline and fuel oil had for a while been hard not to notice, though even in 1974 they had been relatively mild. In 1977 and 1978, Americans underwent harsh winters but surmounted some newsworthy economic dislocations and some discomfort. In 1978 these problems were aggravated by a coal miners' strike, and in 1979 the upheaval in Iran resulted in renewed gasoline shortages. But fossil acreage deprivations were (for a while) nuisances that came and went. People soon adjusted to paying higher prices for fuel after OPEC decreed that they must. When above-ground stockpiles of mined coal were drawn down during the 1978 strike, the United Mine Workers and the coal mine "operators" were jawboned into arriving at a strike settlement, although there was no official or public recognition of the fact that this merely returned them to drawing down exhaustible underground stocks of unmined coal. And if the Europeans thought the North Sea provided the kind of anti-crash insurance they sought, Americans took the same view of Alaska's north slope.

Not visible to the average citizen of the colossal energy-consuming countries was the fact that for several years the ratio of proven oil reserves to on-going consumption had been falling.[2] Not realized by those who now felt reassured that easily usable energy was going to continue to be plentiful was the fact that *several* deposits comparable in magnitude to the North Sea and Alaskan north slope oil fields would need to be discovered *each year* to feed fuel appetites already so colossal and obstinately growing. Instead, what was visible was the fact that the once-controversial pipeline to Valdez was completed. Oil began flowing through it in July, 1977. A few attempts to sabotage it either fizzled or only briefly impeded the flow—and gave Americans the somehow comforting illusion that voracious energy appetites could always be fed as long as adequate "security measures" were taken to protect our hunter-gatherer facilities. Tankers brought the oil from Valdez to California, in sufficient abundance to exceed the capacity of coastal refineries and continental pipelines. The resulting "glut" once more fanned the flames of cornucopianism, giving the myth of limitlessness new plausibility in a time when it ought to have been nearing extinction.

Opportunity Becomes Necessity

In the past, when the myth of limitlessness was more excusably plausible, technological progress ran ahead of population growth and ahead of resource appetite enlargement. This gave us room to grow. When we entered post-exuberant times, we faced a reversal of this situation: growth was continuing and had become a hard-to-break habit, an ingrained expectation now threatening to run ahead of technological advances. Growth momentum was now demanding that the technicians make their breakthroughs as needed to keep us from feeling the pinch of the world's limits. What had been a land of "limitless" opportunity became a land obsessed with conjuring enough new opportunities to keep pace with compulsive growth.

It was not the first time in human history that opportunities had developed into precarious requirements. It is instructive to look at the case of the Irish. There were only about 2,000,000 human inhabitants of the Emerald Isle in 1700, a century after the potato was introduced.[3] At first, potatoes were an addition to the roster of Ireland's food crops; in time they were to become the mainstay of the Irish sustenance base, for reasons I shall not delve into here, having to do with the already troubled history of that country. Eventually they came to be virtually the entire fare of perhaps as many as 90 percent of the people. In 1727, before this had come about, oat crops failed, so the people of Ireland ate up their potatoes about two months sooner than they usually would have done, and many starved before winter had ended. This experience was the provocation for Jonathan Swift's satirical essay (called "A Modest Proposal") depicting a solution to the problems of Ireland by eating superabundant children.

In 1739, Ireland was subjected to an unusual November freezing spell, which destroyed potatoes still in the field and already in storage. It has been estimated that 300,000 people died. But thenceforth, as the Irish became more and more committed to reliance on the potato, their numbers expanded (for a while at an accelerating pace) until they had passed the 8,000,000 mark by the census of 1841. In the top panel of Figure 2, I have plotted the Irish population curve, fitted to dots representing actual data. Population was still rising (probably through a kind of demographic momentum) when a fungus infection of potato plants reached Ireland in 1845. It turned potatoes into stinking, inedible, gooey globs. As the fungus organism (a competitor of *Homo sapiens*) took over the consumption of potatoes, the fungus irruption terminated the prior human irruption.

A people who had become so completely dependent on one type of resource, which now turned out to be a *fallible* resource, were living so precariously that a population crash was inevitable when that resource did indeed fail. Recent censuses have thus shown Ireland (both the Republic and Northern Ireland together) to be inhabited by only about half as many people as lived there just prior to the famine. One striking feature of the Irish experience is the fact that *population was still increasing,* and the rate of increase had barely begun to diminish, *when crash began.* The onset of crash was not heralded by any prior period of "equilibrium."

For about four generations the potato had been an opportunity for the Irish, but it soon became a requirement. For modern prosthetic hunter-gatherer nations, fossil acreage has been similarly converted by exuberant growth (of both population and industry) from an abundance of opportunities into an indispensable requirement. Although in the Irish case we know what happened to population (it crashed, but this crash consisted partly of emigration rather than totally of die-off), we can only surmise the carrying capacity trend. It does seem certain that the 8,175,124 population figure for 1841 had exceeded

FIGURE 2. Omens

Population Experience in Ireland

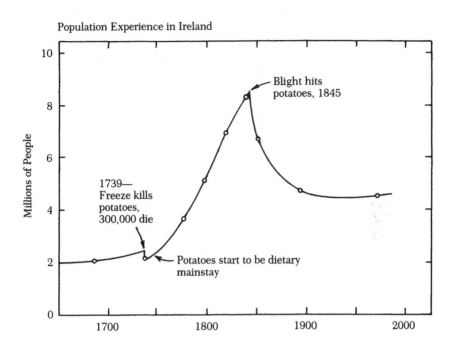

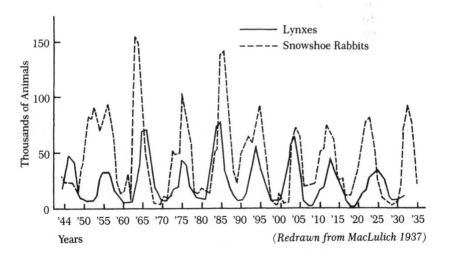

what the sustainable carrying capacity of the land might have been even if potatoes had not been susceptible to blight. Sooner or later soil deterioration from excessive use would have taken a similar toll, though more gradually.

The lower panel of Figure 2 shows the remarkably periodic population cycles of appallingly large amplitude that are possible in a predator-prey pair of species (like the people of Ireland and the potatoes on which they subsisted).[4] The lynx happens to be predisposed to subsist almost exclusively on the snowshoe "rabbit" (the varying hare). Its prey species is capable of rapid multiplication and can thus recover dramatically after its own population crashes. Consumption of rabbits by lynxes appears to be at most only part of the reason for rabbit crashes. Many of the hares die from a syndrome called shock disease, characterized by liver degeneration and terminal hypoglycemia. The incidence of shock disease appears to depend upon population density.[5] (Compare the Sika deer on James Island, mentioned in Chapter 13.) However the malady comes about, as the hare population crashes, the lynxes are left without adequate sustenance, and they crash too. Their too-complete reliance on rabbits is equivalent to the excessive Irish commitment to the potato, and about every ten years the lynxes have undergone the Irish-type experience of relentless increase being abruptly superseded by precipitous decline.

Minds that think ecologically can penetrate beyond the idiosyncrasies of political history, such as the prolongation of feudal warfare in Ireland, and the chronic conflict between the Irish and the English, or between Protestants and Catholics. They can see from the lynx cycles or the Irish population experience the utter folly inherent in the American Petroleum Institute slogan, "A country that runs on oil can't afford to run short." The slogan becomes much truer if we delete its last two words! For any species—human, feline, or whatever—excessive dependence upon a non-constant resource base makes life precarious.

Many former settlements in the American West surged into brief prosperity and then atrophied into ghost towns. Essentially the same process was at work there and in Ireland: people allowed themselves to become numerous and extremely dependent upon a single type of resource that could not remain always abundantly available. Among the ghost towns, of course, "crash" presumably consisted entirely of out-migration; as long as there were other niches in a national economy into which ex-boomtown folk could move, actual die-off was quite unnecessary. Half or more of the Irish crash consisted of emigration. For a *world* population that has overshot carrying capacity, however, emigration is not a feasible alternative to die-off.

Pasts and Their Futures

Even in the face of precedents like Ireland, eminent minds have continued to disregard the prospect of crash for prosthetic societies. Herman Kahn and his associates, for example, assumed that a logistic curve was still the appropriate model to represent the American experience with growth.[6] Moreover, they quite arbitrarily assumed that America had merely arrived at the curve's inflection point by the time of the national bicentennial (see Figure 3, Panel A). That assumption allowed them to project a further 200 years of growth for this nation, albeit progressively decelerating growth. The scale of American life would eventually be twice as high as had already been reached; if actual population increase leveled off much sooner than 200 years hence, then much of the increased scale could take the form of further enlargement of per capita consumption. In any event, there was, they supposed, room for us to become both more numerous and more colossal.

The logistic model hardly applies to all populations. It assumes a constant carrying capacity. It could have applied to Americans if, after European colonization of the North American continent began, the technology used by these new inhabitants had not continued to advance. Then the merely geographic increment of carrying capacity would have produced an expansion of population to fill the newly available but fixed number of niches. At first the expansion would have shown an accelerating pattern. The Age of Exuberance was a manifestation of that pattern, a response to the unexploited niches then available in a New World. Eventually the pattern would have changed to decelerating growth as the ceiling was approached.

Technology did not remain static, however. Recognizing that the technological change that has turned *Homo sapiens* into *Homo colossus* enables us to consume exhaustible resources at escalating per capita rates, and enables us to take over for ecologically unproductive uses land that might otherwise have provided sustenance in the form of renewable resources, Panel B of Figure 3 shows carrying capacity being *reduced* as population expanded in the age of exuberance. The future portion of Panel B depicts what supposedly might happen if, as is advocated by those who have become just enough ecologically

FIGURE 3. Pasts and Their Futures: Four Models

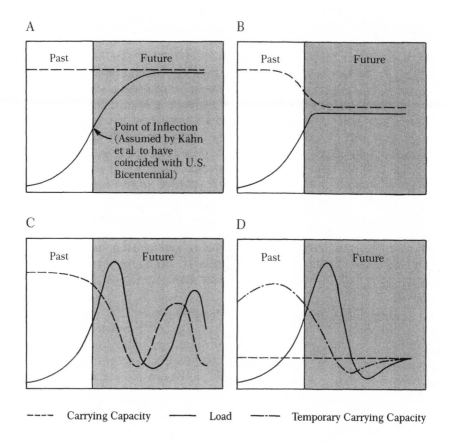

A

| Past | Future |

Point of Inflection
(Assumed by Kahn
et al. to have
coincided with U.S.
Bicentennial)

B

| Past | Future |

C

| Past | Future |

D

| Past | Future |

- - - - Carrying Capacity   ——— Load   —-— Temporary Carrying Capacity

aware to sound the alarm, the nation or the world succeeds in halting
growth very soon. So sharp a turn in the trend curve—from still-ac-
celerating "progress" to a "steady state" almost overnight—would inevitably be one of the most socially traumatic experiences in American
history. Accordingly, those persons who oppose the "steady state society" idea insist that such a transition would be intolerable; many of
them also insist it is not necessary.[7] But even the advocates of such a
revolutionary curbing of aspiration tend to assume that the line between past and future occurs at a point in time where growth has not
yet overshot carrying capacity.[8] If overshoot has already occurred (or
must occur because of demographic and technological momentum already generated), then the model depicted in Panel B is already obso-

lete, whether or not such a transition to equilibrium might once have been feasible.

Resistance to abrupt change of deeply ingrained habits, expectations, and institutional forms and processes probably means the curve could not be voluntarily (or democratically) bent toward the horizontal as sharply as shown in Panel B anyway. Sociocultural (and economic) momentum would tend to make Panel C more realistic. That is, the growth curve will continue upward, even as more and more people begin to sense something like a decline in carrying capacity. Overshoot will occur, if it hasn't already. We may come to feel guilty about stealing from the future, but we will continue to do it. Overshoot will further aggravate the reduction of carrying capacity. Crash must follow. The greater the overshoot, the greater the crash.

After human population has crashed, resource species populations could again grow faster than they would be consumed. Thus, in Panel C, the carrying capacity curve is shown bottoming out and starting to recover as a result of human population crash. Another cycle of human increase follows. This lynx-like recovery would be possible if carrying capacity were entirely or even mainly determined by renewable resources—i.e., if *Homo sapiens* were still a consumer of organic products entirely derived from other (rabbit-like) living components of the biotic community.

This is manifestly not the case. Hence, in Panel D, "carrying capacity" has been represented by two different curves. A major fraction of the recent, apparently high carrying capacity for human high-energy living must be attributed to temporary resources—i.e., non-renewable fossil acreage, the earth's savings deposits. In Panel D, it is optimistically assumed that the component of carrying capacity based on renewable resources has remained stable so far. But it is recognized that serious overshoot, induced by temporarily high composite carrying capacity, will at least temporarily undermine even the sustainable component. "Energy plantations," for example (one of the Cargoist proposals), will tend to aggravate the competitive relation between our fuel-burning prosthetic machinery and ourselves; land taken over to feed technology will not feed humans. So "temporary carrying capacity" is shown actually dipping below the horizontal line for a while, before it recovers and becomes again simply "carrying capacity."[9] The lesson from Panel D is that crash caused by the exhaustion of phantom carrying capacity by *Homo colossus* could preclude a later cycle of regrowth.

The boundary between past and future is drawn in Panel D, as in the other three panels, at a time when population appears not yet to have overshot carrying capacity. Whether or not that optimistic feature of the model is justified by current facts makes little difference, if current practices have committed us to a trajectory that continues upward so that it is destined soon to cross the descending curve that represents global carrying capacity, a capacity not yet acknowledged to be finite. My own view, of course, is that the curves have *already* crossed.

Either way, the past shown in Panel D more nearly accords with ecological history than do the pasts shown in Panels A, B, or C. The future hypothesized by Herman Kahn's think-tank group is dangerously optimistic because it is based on the least realistic past. But the pasts shown in Panels B and C are also less realistic than the past shown in Panel D. The futures shown in Panels B and C are therefore also probably somewhat "optimistic"—although it seems necessary to enclose the word in quotation marks, because even the Panel B future seems dismal, and the Panel C future seems disastrous.

Light from Alaska

The fungus organism that was destroying Ireland's potatoes must have seemed, from 1845 onward, to be the cause of human woe in that land. It would not have been obvious that a more ultimate cause was the fateful deepening of human dependence on a vulnerable resource, and the imprudent human increase induced by that resource's (necessarily temporary) abundance. Similarly, as *Homo colossus* hurtles onward into a future whose central feature has to be a deepening deficit in carrying capacity, the temptation persists to attribute human hardship to such forces as "inflation" that somehow "devours" prosperity. Without an ecological paradigm to remove cornucopian cataracts, people remain blind to resource depletion and see only the rising monetary costs (resulting from pursuit of ever less accessible supplies) as obstacles to the good life.[10] Inexorably escalating prices, rather than physical scarcity, are perceived as the obstacle to continuing lavish consumption and as the source of inequitable distribution. From past habits of thought, tycoons or tyrants are blamed.

But think about Alaska for a moment. It is now a supplier of oil that is imagined to be a kind of insurance against crash. The oil industry spent roughly a thousand times as many dollars to build the pipeline from the north slope to Valdez as the United States government had paid to purchase Alaska! (Even allowing for a century of shrinkage of the dollar, the pipeline cost almost a hundred times as large a fraction of the modern annual GNP as the fraction of the 1867 GNP spent for Alaska.) Many Americans in 1867 had considered exorbitant their government's expenditure of about two cents per acre to acquire "an arctic wasteland." Later, possession of the vast Alaskan wilderness—a final geographic frontier—probably prolonged for some Americans the psychological blessing of the Age of Exuberance. For a few thousand persons each year who migrated northward in recent decades to escape pressures of the colossal way of life that had become standard in the lower 48 states, "Seward's folly" remained until the 1970s a haven. Then, with oil development, there came big money, a deluge of prosthetic apparatus, and opportunists both corporate and individual. "The Great Land" lost much of its haven quality.

Secretary Seward had not quite matched President Jefferson's achievement in territorial expansion, but no doubt we did acquire something valuable in purchasing this northern land. Its greatest value, however, may turn out not to be either the homesteadable land or the oil or other extractable material resources, but the ecological lessons available from research conducted in some of its extreme environments. If we have to be deeply concerned at last about living in a condition of overshoot, we need certain *knowledge* obtainable from Alaska's north slope more than we need its petroleum. Knowledge no one realized would become so humanly salient has been accumulating since the Office of Naval Research began supporting investigations of Arctic ecology through an Arctic Research Laboratory at Barrow, on Alaska's northernmost tip. There, the cyclical blooming and crashing of lemming populations has been one of the phenomena under systematic scrutiny (see top panel in Figure 4). The lemmings live and die according to the kind of pattern shown in Panel C of Figure 3. Research lets us see why.[11]

Lemmings almost disappear every three or four years, after becoming remarkably abundant in between. In their Arctic habitat, biotic communities have much simpler composition (far less species diversity) than in more temperate latitudes. Arctic vegetation tends to consist of dwarf shrubs, grasses, sedges, mosses, lichens, and herbs. All these plants grow very close to the ground. The simplicity of biotic communities in Arctic regions results in periodic population fluctuations that are more easily seen than the less violent and less conspicuously regular cycles occurring in more complex ecosystems.

At Barrow, in seasons when the brown lemmings (*Lemmus trimucronatus*) are abundant, very little vegetation escapes their chopping incisors. Inland, the growing season may be about ten weeks long, but on the coastal plain it is both shorter and more variable,

FIGURE 4. Some Examples of Population Cycles

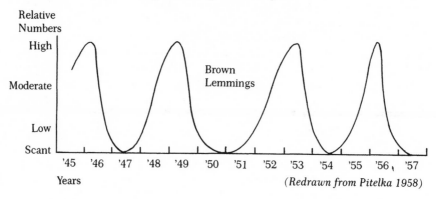

(Redrawn from Pitelka 1958)

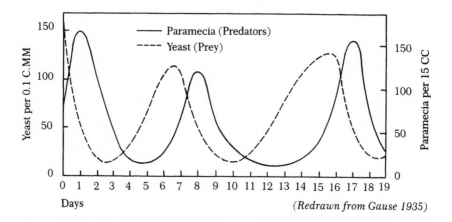

(Redrawn from Gause 1935)

averaging perhaps seven or eight weeks. Lemmings eat all year (beneath the snow cover in winter), consuming accumulated vegetable matter that grew in less than one-fifth of the year. (It is little wonder that their population dynamics resemble the detritovore pattern.) Lemmings reproduce several generations per year.

In habitats where only a single species of lemming is common, a fluctuation that is striking in both amplitude and periodicity results. In peak years, the lemmings may be 500 times as numerous as in the year following a crash and preceding renewed irruption. (It is worth recalling that *Homo sapiens* is now 500 times as numerous as just before agriculture began, and less and less of the world now escapes the hungry jaws of our technology.) Where there are *two* species of small herbivores, population cycles vary. Both the intervals between recognizable periods of abundance and the amplitude of differences between highs and lows may fluctuate. Where there are *more than two* species present, the respective populations will be in competitive interaction. Being differentially susceptible to environmental influences, to predation, etc., the pattern of oscillation for any one population is unlikely to coincide precisely with that of a coexisting species. This affords some stability to the biotic community.

In the Brooks Range foothills, five microtine species are found, with overlapping habitat distributions. There are similarities in their feeding and sheltering activities. Competition among them may depress their respective populations (functioning as an antidote to overshoot) and reduce the likelihood of strong fluctuations. Moving northward toward the coast, the boom-and-bust population cycle in small mammals such as the lemming becomes most clearly evident on the Arctic coastal plains and in lowland areas of regionally extensive marshes. These are the areas where the species diversity of the biotic community becomes most simplified.

Why are the findings of ecological inquiry in the Arctic so valuable to mankind? Because only in regions far beyond the temperate zones do there occur natural ecosystems so simplified by nature's own forces that they provide a clear preview of the destiny toward which mankind is rushing. We rush toward a destiny we have not paused to discern because, under the influence of the cornucopian paradigm, we readily mistake accelerated drawdown (which shortens our future) for a solution to our predicament.

Back to the Takeover Method?

However, it is time to begin seeing that renewed reliance on the older takeover method of enhancing human carrying capacity could be disastrous, too—for a species so numerous and all consuming as humans have become.

With much of the biosphere not yet under human management (though none of it totally escapes human influence anymore), people readily assume that all talk of limits to growth is unnecessary defeatism.[12] Some of the technological powers we have derived from reliance on the drawdown method now enable us to practice the takeover method more effectively than ever. Thus, even if drawdown must eventually fail us, it has armed us to engage again in the older (and supposedly quite benign) method.

Proposals for clearing jungles and turning them into farms and

pastures, making the land support humans instead of jungle creatures, are proposals for renewal of takeover. Proposals for reclaiming
deserts by irrigation, making the land support humans instead of the
creatures adapted by evolution to arid environments, are likewise proposals for further takeover. We know, however, that former forest and
jungle land (especially in the tropics) often turns out to fail as farmland, due to unwanted changes in soil and climate that follow the loss
of tree cover.[13] We know, too, that irrigation leads in time to changes
in soil salinity that have to be resisted by means that didn't used to be
feasible until modern drawdown-based technology was available.

Not all cornucopian takeover proposals envisage taking over land.
Homo colossus even proposes to take over enlarged fractions of the
sunshine upon which the life of this planet depends. To Cargoists,
solutions of energy shortage problems seem feasible partly because
there is so much *unexploited* sunlight falling on every square meter
of the earth's surface. Cargoist minds suppose that all we need to do is
take over for human use (by various devices) solar energy that is now
"going to waste." Proposals of this sort have been offered both by proponents of perpetual economic expansion, and by opponents of other
more conventional but objectionable energy schemes, such as coal
"development" or nuclear power plant construction.[14] Almost never is
consideration given to the possibility that taking over some additional
fraction of the solar flux that now drives the world's climate might
produce ecosystem changes detrimental to human interests. All that
"unused" sunlight is assumed to be "available." (I personally favor
widespread small-scale domestic applications of solar energy technology, especially in preference to most other means of new energy conversion that have been proposed. Nevertheless, I consider it ironic
that solar energy enthusiasts can criticize advocates of coal and fission
for disregarding ecological costs and calculating only monetary costs,
while they themselves glibly regard solar energy as "free" just because
we can't be billed for the incoming sunshine. Nature will have the last
word, and solar enthusiasts have not shown any caution about what it
might turn out to be.)

The trouble with even the takeover method, now that we are so
numerous and have already undergone a revolution of rising aspirations, is that it is like moving down from the Brooks Range foothills
and putting ourselves under the cycle-causing forces characteristic of
the Arctic coastal plain's simplified ecosystem. The more of the earth's
resources we take over for human use (even the renewable ones), the
more nearly we reduce the global ecosystem to one that approximates
the simplistic polarity of the one in which the lemmings live—and in

which they crash. All vegetation serves as sustenance for them, so they are "free" to irrupt and die off, without the damping effect of competitors.

As we saw in Chapter 7, mankind's farms are contrived ecosystems in which tides of natural succession have been turned back. Methodical elimination of all competitors (weeds and pests), and insistence that a given tract of land shall produce a one-species crop, amounts to artificial imposition of a man-and-crops ecosystem type that is as perilously simplified as the lemming-and-tundra type on the Alaskan north coast. Efforts to bring still larger percentages of the world's "arable" land under human cultivation may seem like a solution to problems of scarcity, but with ecological eyes we can see they are an invitation to enhanced probability of disaster. Already we have come a long way toward making the world as ecologically simple as northern Alaska tundra since our ancestors first began sewing and harvesting crops. If we revert to takeover *as a supposed means of sustaining exponential economic growth,* we would not have far to go on any time scale.

The dangers of ecosystem simplification by further takeover are indicated by the lower panel of Figure 4, representing a laboratory experiment.[15] There, the simplest conceivable ecosystem was contrived—two species in a nutrient solution, one species feeding on the other. Carrying capacity for the predator species was synonymous with the abundance of the one other species present, its prey. Functionally, this contrived system simulated the simplicity of the tundra community with the lemmings (or the simplicity of the ecosystem comprising the Irish and their potatoes). If mankind could succeed in eliminating all competitors, and making all coexisting species with which we share "our" world serve as our prey, we would have a world ecosystem essentially similar to the simple one contrived in the lab. If we persisted in denying the possibility of overshoot, would we not, as unwittingly as the paramecium, overconsume our prey, thereby reducing the earth's carrying capacity for us, so that we would necessarily crash, allowing the prey species then to recover, allowing our descendants to irrupt again, and so on?

In effect, the more we might succeed with the takeover method, the more we would thereby be fated to overuse (i.e., draw down) existing populations of renewable resource species. Our success by the drawdown method, using exhaustible resources, has brought us to a situation where some people can seriously contemplate proposals that lead toward almost total takeover. It has become imperative for us to recognize that "success" in such further takeover efforts would require us (like *Paramecium aurelia*, consuming on a faster-than-sustained-yield basis the yeast, *Saccharomyces exiguus*) to revert to drawdown. It was enough of a misstep to have committed ourselves to drawing down non-living resources; it would be a double folly if we were to become any more committed to consuming renewable resources faster than their rate of renewal. *Homo colossus* is already drawing down even the standing crop of timber and other biomass; this is indicated by recent studies showing that the CO_2 content of the world's atmosphere has been increasing over the past century faster than can be accounted for by combustion of fossil fuels.[16] If biomass consumption already exceeds replacement when fossil fuels are still available, what must mankind be expected to do to the global ecosystem as the Carboniferous legacy runs out?

A Third Way: Modesty

Whichever of the two historic approaches we take, either choosing to accelerate drawdown or indulging in additional takeover, our new ecological paradigm enables us to see that eventually we end up shifting back to the other. *Either* traditional way, if prolonged, leads to an inhuman future—not toward the lasting solution of temporarily vexing problems, as Cargoists suppose. For any lasting solution, we must abandon *both* of these ultimately disastrous methods. Drawdown bails us out of present difficulties by shortening our future. Takeover was of lasting value earlier in human history, but that time is past.

We must learn to *live within carrying capacity* without trying to enlarge it. We must *rely on renewable resources consumed no faster than at sustained yield rates*. The last best hope for mankind is ecological modesty.

Red Herring or Rubicon?

Renewal of the takeover method may nevertheless tempt each nation. Some hard ecological thinking will be required to recognize it when it happens, and to discern the ultimate implications. For example, the Tennessee Valley Authority wanted to use a stretch of the Little Tennessee River as a reservoir. Evolution, however, had produced there a small perch species that lived nowhere else. Apparently resistant to transplanting, the snail darter was threatened with extinction by construction of a $116 million dam that would alter its sole habitat beyond its tolerance. Thus mankind and this little fish species became competitors; each had a use for the place that was incompatible with the other's use of it.

Earlier, when Congress had enacted a law requiring human beings and their organized agencies to abstain from extinguishing other species by altering their habitat more than they could tolerate, this had seemed like a good idea. But by the time the Tellico Dam was ready to be put into operation, the federal administration was obliged to deem it silly that a mere fish could preclude any return on the huge monetary investment. So the attorney general argued before the U.S. Supreme Court, with a vial in his hand containing a three-inch specimen of the fish, that this insignificant competitor should not be permitted to cause so much trouble.[17] In June, 1978, however, the Court declared 6 to 3 that it must have been the "plain intent of Congress" when it passed the Endangered Species Act "to halt and reverse the trend toward species extinction, whatever the cost."

The Court's decision prompted congressional action to remove the Act's protection from the particular endangered species that was standing in the way of this piece of technological progress.[18] The way the press portrayed the case was much like the attorney general's presentation. The reading public was thus probably inclined to accept the idea that for a "three-inch fish" to take precedence over a $116 million human investment was ridiculous. Here was an instance where the ecological issue was vastly different from the economic issue, and was far from adequately reflected in even the legal issue. What if an ecologically enlightened Congress had meant to impose a limit on human indulgence in the takeover method—because, carried too far, takeover can lead humanity itself (not just "the tiny fish") into the valley of the shadow of death? And what if the chief justice, who wrote the majority opinion, had been ecologically sophisticated enough to declare that the plain intention of Congress had been just that: to protect the human species by curbing human dominance?

The issue that the attorney general argued was simply whether the snail darter's preservation was worth writing off that $116 million (about 53 cents per American citizen). But the issue that was more important to the human species was whether the time had come to command the forces of takeover to halt and advance no farther.

Wherever the forces of ecological enlightenment might have tried to draw the line, the particular endangered species could probably be made, by legal and economic minds, to seem absurdly insignificant in comparison with human needs. Dolphins, for example, are much more than three inches long and are intelligent mammals; this did not stop the tuna industry from objecting vehemently to marine mammal protective policies that would significantly raise the cost of netting tuna. In light of Figure 4, it should be evident that the question of the significance or insignificance of a particular competitor standing in the way of human expansion may mask the real danger to mankind's future.

Our Best Bet: Expect the Worst

Mankind is condemned to bet on an uncertain future. The stakes have become phenomenally high: affluence, equity, democracy, humane tolerance, peaceful coexistence between nations, races, sects, sexes, parties, all are in jeopardy.[19] Ironically, the less hopeful we assume human prospects to be, the more likely we are to act in ways that will minimize the hardships ahead for our species. Ecological understanding of the human predicament indicates that we live in times when the American habit of responding to a problem by asking "All right, now what do we *do* about it?" must be replaced by a different query that does not assume all problems are soluble: "What must we *avoid* doing to keep from making a bad situation unnecessarily worse?"

Refer again to Figure 3. The most hopeful hypothetical prospect is the one shown in Panel A. But "betting" on that model means going ahead with exuberant drawdown—business more or less as usual. That is probably the worst thing we can "do." That bet will have the effect of falsifying the model ever more drastically (by further hastening the reduction of actual carrying capacity shown in the other three models but not recognized in this one).

If we bet instead on Panel B, supposing there is yet time to effect a swift leveling off before we overshoot the already declining carrying capacity, we may cut our losses in comparison with betting on Panel A. Still, we allow ourselves to assume that resumption of takeover (in place of drawdown) is a safe strategy. By thus letting the snail darter and other inconveniently endangered species become extinct, we move on toward a situation more like Panel C than Panel B.

However, if we chose instead to bet on Panel C, we would be making no distinction between renewable and non-renewable resources, between temporary and permanent carrying capacity, so we would not be taking steps to protect the latter from destruction by technological powers unleashed by the former. This would push us into the situation depicted in Panel D—presumably the worst case.

If we bet on Panel D in the first place, the adaptive steps called for include stringent efforts to protect renewable carrying capacity from both instability arising from excessive (i.e., any further?) take-over and from panic drawdown of already diminishing phantom carrying capacity. Specifically, we would insist on strict enforcement of ecosystem preservation policies prescribed by the Endangered Species Act, the National Environmental Policy Act, and many other pieces of protective legislation dating back to the Antiquities Act of 1906 and beyond. (We would do this for the ultimate sake of our own species.) We would also do our best to stretch out remaining supplies of fossil acreage, instead of competing to hasten their consumption. We would painstakingly revise our cultural values to reduce resource appetites. We would foster non-consumptive modes of human enjoyment, and we would learn to reckon our wealth in terms of our environmental assets rather than in terms of the rate at which we plunder them.

In sum, we would commit ourselves to becoming less colossal, with all deliberate speed. Once we had set that course for ourselves, we would find ourselves committed to more extensive social and cultural change than has ever been achieved (or perhaps even really sought) by past revolutionary movements.[20]

In the final analysis, our best chance of achieving anything like the Panel B future depends on our taking Panel D seriously. Human self-restraint, practiced both individually and especially collectively, is our indispensable hope. It is what Aldo Leopold advocated. It is the essential idea contained in much conservationist legislation. It is the idea that commended itself in an inspired moment to a small band of explorers in the American West in 1870. The men who decided to restrain their own avarice and work for an Act of Congress to preserve intact the wonders of nature they had discovered in their exploration of the Yellowstone area had actually conceived, in their campfire conversation, the key to prolonging human life. They did not know this—nor did anyone else, for a while. But two years of devoted effort did persuade the U.S. Congress to create the world's first national park from the lands this group had explored and discussed. With hindsight, we can see that the Act of Dedication inadequately heralded the significance of a new institutional form. Instead of acknowledging the revolutionary degree of self-restraint implicit in this new land-use concept, the Act merely proclaimed the northwest portion of Wyoming Territory to be a "public park or pleasuring ground" and blandly gave its purpose as "the benefit and enjoyment of the people." A hundred years later there were similar reserves in many lands, dedicated by their respective governments as sanctuaries in which human reactive effects upon local ecosystems were to be held to a minimum, and humans as visitors (rather than inhabitants) could be exposed to a heritage from which they could derive both pleasure and a special awareness of man's part in the biosphere.

An alternative to succession was what the Yellowstone explorers had begun to invent in their campfire discussion. Knowing some of the ways of nature that were reviewed in Chapter 6, it can be seen that in the national park systems of many countries, more than under any other land-use regime, *Homo sapiens* had begun consciously evolving the self-restraint that Aldo Leopold knew was mankind's only alternative to self-destruction.

Decades before we began to realize it, there was an urgent need to begin generalizing to the whole habitat of man, to the whole earth, the concept embodied in that 1870 campfire decision. Not that human societies could, should, or ever would make the whole world into a "public park or pleasuring ground." However, we needed to protect ourselves from succession by somehow protecting our habitat from ourselves.[21]

After a century of experience with national parks, what the nations of the world and mankind at large were still doing to the 99 + percent of the biosphere outside these enclaves required facing up to their most important lesson: namely, that humanity may be best served by an environment we try to avoid changing. Human self-restraint may serve human purposes better than human dominance of the biosphere can. Mankind derives benefits from ecosystems not dominated by man, benefits that may be unavailable from ecosystems man does dominate.

Before the self-restraint embodied in the national park idea was quite one century old, a start *was* made toward generalizing it to the relation of mankind to the entire biosphere. An American manifestation of this belated and faltering start toward generalization was exemplified by congressional passage of the National Environmental Policy Act of 1969 (NEPA). Section 2 said the purposes of the Act were: "To declare a national policy which will encourage productive and enjoyable harmony between man and his environment; to promote efforts which will prevent or eliminate damage to the environment and biosphere and stimulate the health and welfare of man; to enrich the understanding of the ecological systems and natural resources important to the Nation. . . ." But even in the national parks, harmony between man and environment was proving to be an elusive goal. The volume of visitation became excessive, and people impaired by overuse the very specimens of wild nature they sought explicitly to preserve.[22]

American congressmen in 1969 were too new to ecological thinking to realize that what they were trying to do when they enacted NEPA was to halt by legislative command the man-caused succession overtaking man. They certainly had little comprehension of succession's inexorability. NEPA required, among other things, that every recommendation for future legislation or other federal action "significantly affecting the quality of the human environment" must include a detailed statement on the environmental impact of the proposed action, particularly its adverse environmental effects, alternatives to it, and the long-term resource commitments it would entail. Ostensibly, projects that would result in too much destructive environmental impact were to be abandoned after such an environmental impact statement (EIS) making this evident had been reviewed by higher authority. But this was only implicit, and before long the filing of EIS's tended to become defined as a pro forma nuisance, after which habitat exploitation as usual was to remain the norm.

But we had tried, and this needs to be remembered. Three years later, on a wider scale, the *world* tried again to generalize the 1870 norm of self-restraint, when the United Nations Conference on the Human Environment met in Stockholm.[23]

It may be worth asking, what if the 41st Congress (instead of the 91st) had enacted NEPA in 1869 (instead of in 1969)? What superhuman imagination would the 39 million post–Civil War Americans have needed to be able to decide then that their descendants a century later would be better served by their environment if they did not become five times more numerous and twenty times more urbanized!

What inconceivable foresight would have been required for the members of the Virginia House of Burgesses (meeting in Williamsburg in 1769) to decide for their constituents' descendants over the next two centuries that the western lands then claimed by that English colony should be protected from eventual industrial exploitation! Could even a mind like Jefferson's have discerned the need to require an EIS to be filed by anyone who might invent strip-mining as a means of obtaining massive quantities of coal from under those lands? Could anyone in the colonial capital have been expected to visualize eight generations ahead of time (when prevention might have been painless) the huge prosthetic machines used by the Central Ohio Coal Company for digging into the portions of Virginia's western claim that would long since have become the state of Ohio?

265

What if Ferdinand and Isabella had been cautious and had required Christopher Columbus to submit an EIS before they authorized his proposed project of exploration? What could the earnest navigator have foreseen as the consequences likely to ensue from his voyage of discovery? Had he submitted an impact statement, could it possibly have described a forthcoming Age of Exuberance, the myths it would foster, and the democratic institutions it would nurture? Would the impact statement have predicted the irruption of *Homo sapiens* to fill up the carrying capacity surplus the voyage was going to disclose, or the Cargoism, Cosmeticism, Cynicism, and Ostrichism by which post-exuberant men would strive to perpetuate obsolescent ways? Would the explorer's royal patrons have believed such wild ideas? Would they have sought ways to minimize the regrettable portions of the project's probable impact?

No humans ever had that kind of imagination and foresight. We know, with hindsight, what our predecessors did not and could not foresee. By seeing with an ecological paradigm, we can recognize that filing essentially conjectural EIS's and then trying somehow to walk softly (but in a still exploitative approach to the future) will not suffice. A much more basic change of direction is called for. We must become less detritovorous. We must maintain symbiosis with competitors to keep from falling into the pattern of the lemmings in Alaska, or the paramecia in the laboratory.

Our best bet is to act as if we believed we have already overshot, and do our best to ensure that the inevitable crash consists as little as possible of outright die-off of *Homo sapiens*. Instead, it should consist as far as possible of the chosen abandonment of those seductive values characteristic of *Homo colossus*. Indeed, renunciation of such values may be the main alternative to renewed indulgence in cruel genocide. If crash should prove to be avoidable after all, a global strategy of trying to moderate expected crash is the strategy most likely to avert it.

History will record the period of global dominance by *Homo colossus* as a brief interlude. Our most urgent task is to develop policies designed not to prolong that dominance, but to ensure that the successor to *Homo colossus* will be, after all, *Homo sapiens*. Developing such policies must be so enormously difficult that it is not easy even to accept the urgency of the task. But the longer we delay beginning, the more numerous and colossal we become—thereby trapping ourselves all the more irredeemably in the fatal practice of stealing from our future.

Notes

1. For comparison of "tech fixes" with alternatives in the context of a problem other than energy, see T. A. Heberlein, "The Three Fixes: Technological, Cognitive, and Structural," pp. 279–296 in Don R. Field, J. C. Barron, and B. F. Long, eds., *Water and Community Development: Social and Economic Perspectives* (Ann Arbor: Ann Arbor Science Publishers, 1974).

2. See Hubbert 1969 (listed among references for Ch. 10); Freeman 1974 (listed among references for Ch. 14), pp. 87–91. Not just average citizens, but scholars too remained preoccupied with the precariousness of depending on *imported* oil rather than concerning themselves about the precariousness of depending on *exhaustible* oil. See Robert Stobaugh, "After the Peak: The Threat of Imported Oil," Ch. 2 in Stobaugh and Yergin 1979 (listed among references for Ch. 14).

3. Most of the population data and much of the Irish history described in these paragraphs and portrayed in Figure 2 are based on the very illuminating account in Salaman 1949; see also Woodham-Smith 1962, and Kammeyer 1976.

4. For analyses of these cycles, see MacLulich 1937; Elton and Nicholson 1942.

5. See Green et al. 1939.

6. Kahn et al. 1976.

7. For example, see Neuhaus 1971 (listed among references for Ch. 11); see also John R. Maddox, *The Doomsday Syndrome* (New York: McGraw-Hill, 1972), and Irving Louis Horowitz, "The Environmental Cleavage: Social Ecology versus Political Economy," *Social Theory and Practice* 2 (Spring, 1972): 125–134; H. S. D. Cole, Christopher Freeman, Marie Jahoda, and K. L. R. Pavitt, eds., *Models of Doom: A Critique of the Limits to Growth* (New York: Universe Books, 1973).

8. For example, see Herman E. Daly, *Steady State Economics* (San Francisco: W. H. Freeman, 1977); Mancur Olson and Hans H. Landsberg, *The No-Growth Society* (London: Woburn Press, 1975); Dennis C. Pirages, ed., *The Sustainable Society* (New York: Frederick A. Praeger, 1977).

9. Although, as noted in Ch. 2, note 12, "temporary carrying capacity" is a contradiction in terms, here the phrase is used to designate a temporarily enlarged composite consisting of true (i.e., sustainable) carrying capacity plus phantom carrying capacity.

10. For an exception, indicating that even in economic analysis the "cataracts" may have begun to be removed, see Brian L. Scarfe, *Cycles, Growth, and Inflation: A Survey of Contemporary Macrodynamics* (New York: McGraw-Hill, 1977), pp. 289–299.

11. See Pitelka 1958.

12. See note 7.

13. Ehrlich and Ehrlich 1972 (listed among references for Ch. 12), pp. 202–204.

14. Lovins 1977 is obviously opposed to further expansion of centralized high technology energy systems, and favors rapid development of decentralized systems using renewable energy sources—i.e., solar energy, directly or indirectly. On the other hand, the Energy Project at the Harvard Business School, which also advocates low-technology solar energy development, shows its commitment to perpetual expansion when it asks us to regard energy *conservation* as "an alternative energy source." See Daniel Yergin, "Conservation: The Key Energy Source," Ch. 6 in Stobaugh and Yergin 1979 (listed among references for Ch. 14). Referring to conservation as a "source" of energy makes sense only if it is assumed that the purpose in saving energy from present uses is to divert it to other uses. If instead our goal is to reduce ecological load so it no longer exceeds carrying capacity, then conservation is not a "source" of energy but an antidote to overshoot.

15. Gause 1935; cf. Utido 1957.

16. Woodwell 1978.

17. "Bell Urges Court to Permit Dam to Open Despite Peril to Rare Fish," *New York Times,* Apr. 19, 1978, p. A19.

18. Although subsequently the dam's threat to local farmers whose land was to be submerged and to Indian tribes whose ancient burial grounds would be inundated aroused further resistance, the pro-technology, pro-takeover forces seemed to prevail after President Carter decided not to veto the legislation exempting Tellico Dam from the Endangered Species Act. For reaction to that decision, see Luther J. Carter, "Carter's Tellico Decision Offends Environmentalists," *Science* 206 (Oct. 12, 1979): 202–203.

19. The tendency persists, however, to view the jeopardy in altogether political terms, unmindful of its ecological foundation. See, for example, Moss 1975; cf. Buultjens 1978.

20. Cf. Tudge 1977; Steinhart and Steinhart 1974, especially Ch. 12, "Return to the Future."

21. As was pointed out in Ch. 10, bloom and crash constitute a special kind of sere; crash is a kind of "succession without a successor" where there happens to be no pre-adapted replacement species available to occupy a new niche made available by the environmental alterations that evicted the crashed species.

22. Darling and Eichhorn 1967.

23. For a revealing perspective on the relation between the Stockholm conference on the environment and the subsequent Rome conference on food, see Thomas G. Weiss and Robert S. Jordan, *The World Food Conference and Global Problem Solving* (New York: Frederick A. Praeger, 1976), pp. 138–142.

Selected References

Buultjens, Ralph
    1978. *The Decline of Democracy: Essays on an Endangered Political Species.* Maryknoll, N.Y.: Orbis Books.
Darling, F. Fraser, and Noel D. Eichhorn
    1967. *Man and Nature in the National Parks: Reflections on Policy.* Washington: Conservation Foundation.
Elton, Charles, and Mary Nicholson
    1942. "The Ten-Year Cycle in Numbers of the Lynx in Canada." *Journal of Animal Ecology* 11 (Nov.): 215–244.
Gause, G. F.
    1935. "Experimental Demonstration of Volterra's Periodic Oscillation in the Numbers of Animals." *Journal of Experimental Biology* 12 (Jan.): 44–48.
Green, R. G., C. L. Larson, and J. F. Bell
    1939. "Shock Disease as the Cause of the Periodic Decimation of the Snowshoe Hare." *American Journal of Hygiene* 30 (Nov.): sec. B, pp. 83–102.
Kahn, Herman, William Brown, and Leon Martel
    1976. *The Next 200 Years: A Scenario for America and the World.* New York: William Morrow.
Kammeyer, Kenneth C. W.
    1976. "The Dynamics of Population." Pp. 189–223 in Harold Orel, ed., *Irish History and Culture: Aspects of a People's Heritage.* Lawrence: University Press of Kansas.
Lovins, Amory B.
    1977. *Soft Energy Paths: Toward a Durable Peace.* New York: Harper & Row.
MacLulich, D. A.
    1937. "Fluctuations in the Numbers of the Varying Hare (*Lepus americanus*)." *University of Toronto Studies,* Biological Series no. 43: 5–136.
Moss, Robert
    1975. *The Collapse of Democracy.* London: Temple Smith.
Pitelka, Frank A.
    1958. "Some Characteristics of Microtine Cycles in the Arctic." Pp. 73–88 in Henry P. Hansen, ed., *Arctic Biology.* Corvallis: Oregon State College, Eighteenth Annual Biological Colloquium.
Salaman, Radcliffe Nathan
    1949. *The History and Social Influence of the Potato.* Cambridge: Cambridge University Press.
Steinhart, Carol, and John Steinhart
    1974. *The Fires of Culture: Energy Yesterday and Tomorrow.* North Scituate, Mass.: Duxbury Press.
Tudge, Colin
    1977. *The Famine Business.* London: Faber and Faber.

Utido, Syunro
1957. "Cyclic Fluctuations of Population Density Intrinsic to the Host Parasite System." *Ecology* 38 (July): 442–449.
Woodham-Smith, Cecil
1962. *The Great Hunger: Ireland 1845–1849*. New York: Harper & Row.
Woodwell, George M.
1978. "The Carbon Dioxide Question." *Scientific American* 238 (Jan.): 34–43.

Glossary

ABIOTIC SUBSTANCE: the non-living elements and compounds of the environment.

ADAPTATION: modification of something to "fit" or "go with" something else; hence, in ecology, the fitting of an organism to the requirements of its environment by means of adjustments of form or function.

AGE OF EXUBERANCE: the centuries of growth and progress that followed the sudden enlargement of habitat available to Europeans as a result of voyages of discovery; a period of expansion when a species takes exuberant advantage of the abundant opportunities in an eminently suitable but previously inaccessible habitat.

ANIMAL COMMUNITY: a biotic community in which animals play a conspicuous part in the collective adaptation of the total association of diverse organisms to their shared habitat.

ANOMIE: prevalence of the view that antisocial types of behavior are necessary as means to attain socially prescribed goals.

ANTAGONISM, ECOLOGICAL: a relationship between two or more species or groups of organisms in which the activities or indirect effects of the activities of one are harmful to another; may be quite impersonal, unintentional, and unconscious.

ANTAGONISM, EMOTIONAL: hostility, enmity, active opposition toward another individual or group; may result from ecological antagonism.

ANTIBIOTIC: (adj.) opposite of symbiotic; tendency for the life processes and products of one organism to be harmful to another organism or type of organism; (n.) chemically purified extrametabolite used in medicine to inhibit growth of disease-causing bacteria—e.g., penicillin, refined from an extrametabolite exuded by the penicillium mold.

ANTINOMIAN MOVEMENTS: groups of people who seek to change social conditions they deplore by repudiating or violating social or legal codes which they assume have caused the deplored conditions.

AUTOTROPHS (producers): organisms capable of photosynthesis, which converts abiotic substances into energy-rich organic substances. (Also, technically, organisms capable of chemosynthesis, whereby they obtain energy to build up organic substances by oxidizing inorganic compounds; not directly important to the argument of this book, so herein the word is used to denote only those organisms capable of photosynthesis.)

BIOGEOCHEMICAL CYCLES: patterns of circulation of energy and a score of chemical elements involved in life processes, in systems comprising earth, water, air, and organisms. The chemical materials can be and are recycled over and over again; the energy cannot be recycled.

BIOMASS: the amount of living material in a specified context.

BIOSPHERE: the region at and near the earth's surface within which life occurs, including land, water, and air; all life on the entire planet.

BIOTIC COMMUNITY: a more or less self-sufficient and localized web of life collectively adapting to the life-supporting conditions of the local habitat.

BIOTIC POTENTIAL: the capacity for reproduction, or the number of offspring that theoretically could be produced by a parental pair.

BLOOM: *see* irruption.

CARGO CULTS: systems of belief and ritual among Melanesian peoples in the islands of the Pacific, arising from their observation that the dominant European minority among them received vast quantities of material goods from unseen sources overseas; the native peoples, because of their sketchy and inaccurate notions of the origins of these goods, came to suppose they might be the rightful owners of such cargo and that it might ultimately be delivered to them.

CARGOISM: faith that technological progress will stave off major institutional change even in a post-exuberant world; the equivalent among people of industrial nations to the cargo cults of the Melanesian islanders.

CARNIVORE: an animal type that subsists by eating the flesh of other animals.

CARRYING CAPACITY: the maximum population of a given species which a particular habitat can support indefinitely (under specified technology and organization, in the case of the human species).

CARRYING CAPACITY DEFICIT: the condition wherein the permanent ability of a given habitat to support a given form of life is less than the quantity of that form already in existence.

CARRYING CAPACITY SURPLUS: the condition wherein the permanent ability of a given habitat to support a given form of life exceeds the quantity of that form then in existence.

CLIMAX COMMUNITY: a self-perpetuating community comprising a combination of species that can successfully outcompete any alternative combination which might otherwise replace it on a given site.

COLOSSUS: originally the gigantic statue of Apollo at the entrance of the harbor of Rhodes; hence, any gigantic person or thing; in this book, a human being equipped with tools or apparatus that greatly enlarge the resource demands and environmental impact of that organism.

COMPETITION: a relationship between two or more organisms in which each makes demands upon its environment similar to the demands made by the other, with the result that the presence of each hampers the other in some way.

CORNUCOPIAN PARADIGM: a view of past and future human progress that disregards the carrying capacity concept, pays no attention to the finiteness of the world or to differences between takeover and drawdown, and accepts uncritically the myth of limitlessness.

COSMETICISM: faith that relatively superficial adjustments in our activities will keep the New World new and will perpetuate the Age of Exuberance.

CRASH: the more or less precipitate decline in numbers that follows when a population has exceeded the carrying capacity of its habitat; otherwise called a die-off.

CULTURAL LAG: stress that occurs when interconnected patterns of culturally prescribed behavior or belief change at different rates of speed.

CULTURE: a system of socially acquired and transmitted standards of judgment, belief, and conduct, as well as the social and material products of the resulting conventional activities.

CULTURE OF EXUBERANCE: the total complex of beliefs and practices associated with the opportunities for expansive life in the Age of Exuberance; a culture founded upon the myth of limitlessness.

CYNICISM: ordinarily, a tendency to question the value of anything and everything; hence, in this book, disbelief that the New World's newness made any difference or that its oldness has any significance; disbelief that the Age of Exuberance was of any value, or that its termination matters.

DECOMPOSERS: heterotrophic organisms (usually microscopic) that break down the complex organic compounds of dead plants or animals, thereby releasing simple substances that can be re-used by autotrophic producer organisms.

DETRITOVORE: an organism that subsists by consuming detritus; by extension, any organism that uses the accumulated remains of long-dead organisms, including industrial human communities which are "detritovorous" insofar as they depend on massive consumption of the transformed organic remains from the Carboniferous period known as fossil fuels.

DETRITUS: originally a geological term meaning fragments of rock or debris produced by disintegration and erosion; used by biologists to refer to the accumulated remains of organisms (e.g., humus in the soil, or decayed leaves in a pond); by extension, used in this book to refer to transformed remains of organisms that lived millions of years ago, such remains being useful as fossil fuels to organisms (humans) living today.

DETRITUS ECOSYSTEM: an ecosystem in which detritovores play a major part. As organic detritus accumulates in a given habitat, there is a temporary increase in carrying capacity for detritus consumers. Insofar as these are capable of increasing much faster than the detritus accumulates, however, their introduction to the community after detritus has already accumulated, or their release from some constraint that had earlier held back their use of the accumulation, tends to result in a cycle of bloom and crash. They irrupt and then, as the detritus supply is exhausted, they die off.

DEUS EX MACHINA: in ancient Greek drama, a deity brought in by stage machinery to intervene in the action; hence, any artificial or improbable resolution of a dilemma. In this book, Cargoism is regarded as an instance of this. (*See* Cargoism.)

DIACHRONIC COMPETITION: a relationship between generations in which living organisms satisfy their wants at the expense of their descendants.

DOMINANCE: the capacity of a species to exert more influence than any other associated species in its community upon the characteristics of their habitat and upon other species in the community; ecological inequalities that arise from different resource-exploitation strategies characterizing different types of organisms.

DOMINANT: (adj.) having the capacity to exert more influence than any other species in the community; (n.) the particular species in a community with greater influence than any other.

DRAWDOWN METHOD OF EXTENDING CARRYING CAPACITY: an inherently temporary expedient by which life opportunities for a species are temporarily increased by extracting from the environment for use by that species some significant fraction of an accumulated resource that is not being replaced as fast as it is drawn down. (*See,* by contrast, Sustained yield.)

ECOLOGICAL PARADIGM: in general, a view of the web of life that recognizes a common chemical basis for all types of organisms (including man), emphasizes the dependence of all life processes upon flows of energy and exchanges of chemical substance between organism and environment, and expects living forms inevitably to have effects upon each other by these exchanges; in this book, rejection of the notion of human exemption from ecological principles and affirmation of the view that ecological concepts are essential for understanding human experience.

ECOLOGY: the study of ecosystems; i.e., the study of interrelations in the web of life and between organisms and their environment.

ECOSYSTEM: a comprehensive web of interrelations between organisms, other organisms, and their environment; it tends to be characterized by the operation of various checks and balances.

ENCLOSE-AND-CONTROL PRINCIPLE: any adaptation wherein organisms retain homeostatically within themselves conditions necessary to their life that have ceased to be reliably available externally; the principle by which, for example, organisms that were once confined to aquatic environments became adapted to life on dry land by maintaining within themselves a wet "internal environment."

ENERGY SLAVES: a metaphor by which the value of fossil energy is measured in terms of the human muscle-power equivalent to it. For example, burning about 32 U.S. gallons (121 liters) of gasoline releases energy equal to the energy content of the food an active adult human being would consume in a year. Thus, the work obtained from this gasoline tends to approximate the amount of work that might have been obtained in a year from a human slave, assuming comparable energy conversion efficiency by fuel-burning machinery and food-burning human bodies.

ENERGY SUBSIDY: energy from sources other than sunlight applied to the growing of crops; e.g., fossil energy (in excess of the human labor displaced by its use) used in operating farm equipment, energy used to move water for artificial irrigation, the energy content of synthetic fertilizers applied to the soil, etc.

ETHNOCENTRISM: the tendency to regard the behavior of people raised by other standards than our own as wrong, rather than just different.

EVOLUTION: imperfect replication of the traits of organisms in their progeny and the selective retention among descendant populations of those traits best adapted to prevailing environmental circumstances.

EXTRAMETABOLITES: substances given off by organisms which, as they accumulate in the environment, affect the life processes of other organisms, and thus function as environmental chemical regulators.

EXUBERANCE, ECOLOGICAL: the lavish use of resources by members of a freely expanding population who are, at a given time, significantly fewer in number than the maximum permitted by the carrying capacity of their habitat.

EXUBERANCE, EMOTIONAL: a joyous, optimistic, almost euphoric mood; can result from ecological exuberance.

EXHAUSTIBLE RESOURCES: usable substances that are not currently being re-created by natural processes at rates comparable to actual or potential rates of consumption; e.g., detritus, fossil fuels, minerals, metallic ores.

FATE: in human experience, a future that happens to us regardless of our own actions; as defined by the sociologist C. Wright Mills, the summary outcome not intended by anyone but resulting from innumerable small decisions about other matters by innumerable people. For example, no one intended that city air should become smoggy, but the cumulative effects of individual decisions to buy and drive automobiles, to burn fuels to heat homes, or to buy products of fuel-using industries have brought about this unintended environmental change.

FISH ACREAGE: the additional farmland a given nation would need in order to supply food equivalent to that portion of its sustenance obtained from the sea.

FOOD CHAIN: a hierarchy of relationships between species that consume or are consumed by other species; e.g., grass eaten by cow, beef or dairy products eaten by humans, human blood consumed by mosquitoes, mosquitoes eaten by trout, etc.

FOSSIL ACREAGE: the additional farmland a given nation would need in order to supply organic fuel equivalent to the coal, petroleum, or natural gas products it now uses.

FOSSIL FUELS: energy-rich substances created by geological transformation of the remains of organisms that lived long ago; including coal, petroleum, and natural gas.

GENERATION: as a measure of time, the average interval between becoming a parent and becoming a grandparent, or between being born and becoming a parent; assumed herein to be about 25 years for *Homo sapiens*.

GHOST ACREAGE: the additional farmland a given nation would need in order to supply that net portion of the food or fuel it uses but does not obtain from contemporary growth of organisms within its borders—e.g., from net imports of agricultural products, from oceanic fisheries, from fossil fuels.

GREEN REVOLUTION: a dramatic increase in agricultural productivity resulting from use of specially bred crop plants capable of higher per-acre yields than varieties previously in use.

HERBIVORE: an animal type that subsists by eating plant tissues.

HETEROTROPH (consumer): an organism which cannot convert abiotic substances into organic matter but must use for its sustenance materials produced by other organisms.

HOMEOSTASIS: the prevention of change in some variable by means of compensatory change in another aspect of the system when an outside influence would otherwise have produced change in that variable; the damping of oscillations in biological processes by regulatory mechanisms.

HOMO COLOSSUS: *See* Colossus.

HOMO SAPIENS: the species of mammal that includes both the reader and the author, and all other contemporary human beings; a language-using, toolmaking, social species, descended from earlier types of humans who were also members of the genus *Homo,* capable of evolving culturally as well as genetically.

HUMAN COMMUNITY: a biotic community in which human beings, by virtue of their capacity for occupational and social differentiation, play so many of the roles that it appears the community is mostly under human control.

HUMAN ECOLOGY: properly defined, the comprehensive study of ecosystems involving mankind. (Now a thoroughly interdisciplinary field of inquiry, among sociologists human ecology was regarded as a branch of academic sociology; they tended for a while to narrow it to the study of little more than the spatial structure of cities, afterward broadening it again to a more general study of social organization.)

HUMAN EXEMPTIONALISM: the notion that human beings are so fundamentally unlike other living creatures that principles of ecology (and perhaps many of the principles of other branches of biology, too) are inapplicable to us.

IRRUPTION: the rapid exponential increase of a population after it suddenly gains access to an abundance of the resources it requires.

KILOCALORIE: a unit of energy; the amount of heat required to raise the temperature of 1 kilogram of water 1 degree Celsius.

LIMITING FACTOR: any condition in an ecosystem that tends to restrict growth of a population of organisms (or, the resource a given population needs that happens to be available in lowest per capita abundance).

METABOLISM: physical and chemical processes occurring in living cells, organisms, and biotic communities, whereby compounds are either built up or broken down, and energy is either fixed or released. (Includes both anabolism, the process by which food is turned into living tissue, and catabolism, the process by which tissue is turned into waste products.)

MUTUALISM: a strong and reciprocal interdependence between different but associated life forms. (*See* symbiosis.)

MYTH OF LIMITLESSNESS: the belief (more implicit than explicit, perhaps) that the world's resources are sufficient to support any conceivable human population engaged in any conceivable way of life for any conceivable duration; derivatively, the belief either that a given resource is inexhaustible or that substitutes can always be found.

NEO-EXUBERANT: newly committed to the culture of exuberance (whether or not the carrying capacity surplus that gave rise to that culture still exists).

NET PRODUCTION: the quantity of organic matter produced during a given time period by photosynthesis in a given organism (or community) minus the quantity destroyed by respiration.

NEW REALITY, THE: the situation prevailing in the world following transition from a carrying capacity surplus to a carrying capacity deficit.

NICHE: the role that an organism (of a given kind) plays in an ecosystem—i.e., the kinds of nutrients it has to obtain from its environment, the kinds of things it must do to its environment in the process of living, the kinds of relationships it must have with other organisms to go on living, etc.

NICHE DIVERSIFICATION: increase in the structural complexity of a biotic community through enlargement of the number of distinct ecological niches; a common result of population pressure in a given niche. Among humans, occupational diversification is a special case of this natural process.

OSTRICHISM: obstinately persistent belief in the myth of limitlessness; the unrealistic supposition that nothing basic has changed; refusal to face facts.

OVERPOPULATION: population in excess of carrying capacity; population so numerous in proportion to resources that standards of living are lower than they would be if population were less numerous.

OVERSHOOT: (v.) to increase in numbers so much that the habitat's carrying capacity is exceeded by the ecological load, which must in time decrease accordingly; (n.) the condition of having exceeded for the time being the permanent carrying capacity of the habitat.

PARADIGM: an underlying shared idea of the fundamental nature of whatever it is that a collectivity of minds seeks to understand; an idea that guides inquiry and thought by defining what seems real, how things are presumed to work, and how additional facts about this reality and these processes may presumably be obtained.

PHANTOM CARRYING CAPACITY: illusory or extremely precarious capacity of an environment to support a life form or a way of life; that portion of a population that cannot be permanently supported when temporarily available resources become unavailable.

PHOTOSYNTHESIS: creation of organic molecules from abiotic substances by the action of solar energy (a process that occurs in green plants, for example, by means of the catalytic action of a green substance called chlorophyll).

PIONEER COMMUNITY: the type of association of organisms that is capable of using a site early in its developmental history; the earliest pre-climax community on a given site.

PLANT COMMUNITY: a biotic community in which the dominant species are plants.

POLLUTION: the accumulation of harmful substances produced by human activity in the environment in which humans live; by extension, the accumulation of any life-limiting extrametabolites in the environment of any species.

POPULATION DENSITY: ratio of number of organisms of a given species to number of units of space in which they live; e.g., people per square mile.

POPULATION PRESSURE: the frequency of mutual interference per capita per day resulting from the presence of others in a finite habitat.

POST-EXUBERANT: the condition of life in a world that has gone through an Age of Exuberance but now confronts a carrying capacity deficit.

PRE-CLIMAX COMMUNITY: any community subject to succession because its collective use of a site is not self-compensating and the characteristics of the site are changed by that community's life processes so the site becomes less suited to supporting the existing association of organisms and more suited to supporting a changed association.

PRE-ECOLOGICAL THOUGHTWAYS: *see* cornucopian paradigm.

PRODUCTION: in ecology, the fixation of energy by photosynthesis (or, the construction of energy-rich organic matter from energy-poor abiotic substances, by photosynthetic action).

PROSTHETIC DEVICE: in medical practice, an artificial substitute for a part of the body (as, for example, an artificial limb); by extension, any artificial device or any other thing that either serves a function some organ would otherwise serve, or enables an organism to do something it could not otherwise do without having developed a special organ for the purpose.

QUASI-SPECIATION: the non-genetic differentiation of a human population into differently specialized subgroups by use of alternative tools, customs, or symbols.

REALISM: recognition that the Age of Exuberance is over and that overpopulation and resource depletion must inexorably change human organization and human behavior.

RECYCLING: the natural return of substances from the end of a food chain to the beginning, through the action of decomposers; by extension, artificial processes in which previously used materials are reused.

REDUNDANCY ANXIETY: a morbid apprehension arising from population pressure, based on the more or less conscious realization that if there is an excess of population in relation to carrying capacity, the population surplus may include oneself, not just others.

RENEWABLE RESOURCES: usable substances produced by on-going processes such as organic growth that takes place at rates commensurate with actual or potential rates of consumption; also usable energy obtained directly or indirectly from contemporary solar inputs, rather than withdrawn from finite quantities accumulated from past solar inputs.

RESOURCES: materials or energy sources usable by organisms. Not only the characteristics of the materials or sources, but also the traits of the organism and its equipment, will determine which materials or sources are usable.

RESPIRATION: chemical reactions in organisms by which energy in organic molecules is made available for use, mainly through oxidation of carbon compounds.

SELECTION PRESSURE: the favoring of one variant over another by its more advantageous adaptation to environmental circumstances; measured by differential rates of replacement.

SERAL STAGES: the developmental stages in a process of community succession.

SERE: the entire sequence of community types that tend to succeed each other on a given type of site.

SPECIATION: differentiation of a population into distinct types that do not interbreed and that have genetically produced traits adapted to different environmental conditions.

SPECIES: a category of organisms taxonomists have judged to be sufficiently distinct from others for recognition as a separate kind; assumed (or known) to be incapable of interbreeding with another species.

STANDING CROP: *see* biomass.

SUBDOMINANT: (n.) a species with significant but less than first-rank influence upon the characteristics of its habitat and upon other species sharing that habitat; (adj.) having less than first-rank influence.

SUBSPECIES: a partially differentiated sub-population still capable of inter-breeding with other sub-populations of the same species; e.g., a race.

SUCCESSION: an orderly and directional process of community change resulting from modification of the habitat by the community and culminating in a maximally stable ecosystem relative to the characteristics of the site.

SUSTAINED YIELD: the result of balance between rates of harvest and rates of growth and replacement, so that the resource is not destroyed by overuse.

SYMBIOSIS: mutual dependence of populations of differentiated organisms.

SYMBIOTIC: mutually interdependent, especially in the positive sense of mutually beneficial.

TAKEOFF: in economics, the point in the history of a society at which its economic surplus is sufficient to permit continual reinvestment in economic growth, so that growth becomes self-sustaining.

TAKEOVER METHOD OF ENLARGING CARRYING CAPACITY: increasing opportunities for one species by reducing opportunities for competing species.

TECHNOLOGY: either the systematic study and knowledge of the means of making and using devices for implementing the attainment of human goals, or the devices themselves.

TEMPORARY CARRYING CAPACITY: combination of actual carrying capacity and phantom carrying capacity; the population that a habitat can support for a short time only (until the supply of some exhaustible resource upon which that species has become dependent runs out).

TERRITORIALITY: an animal behavior pattern that commonly arises in response to actual or potential resource shortages: individuals or groups lay claim to distinct territories for feeding or breeding, drive off non-claimants, and thereby ensure that scarce resources will adequately support claimants rather than being spread too thinly among an excessive population.

TRADE ACREAGE: the additional farmland a given nation would need in order to supply that net portion of food or fuel it obtains by trade with other nations.

VILIFICATION: reviling or defaming other individuals or groups; imputing to supposed villains major responsibility for evil developments.

VISIBLE ACREAGE: the farmland, forests, and pastures within a nation's own borders which it recognizes as the source of products it consumes, or as the source of commodities it trades for imports it consumes.

WEB OF LIFE: the total network of cooperative, competitive, and predatory relationships between organisms struggling for existence in a common and finite habitat.

Indexes

Name Index

Subject Index

UNIVERSITY OF ILLINOIS PRESS
1325 SOUTH OAK STREET
CHAMPAIGN, ILLINOIS 61820-6903
WWW.PRESS.UILLINOIS.EDU